Ordinary Differential Equations

HARPERCOLLINS COLLEGE OUTLINE

Ordinary Differential Equations

Paul Duchateau, Ph.D.
Colorado State University

HarperPerennial
A Division of HarperCollins*Publishers*

An American BookWorks Corporation Production

Project Manager: William R. Hamill

Editor: Robert A. Weinstein

Library of Congress Catalog Card Number: 91-55403

ISBN: 0-06-467133-X

92 93 94 95 96 ABW/RRD 10 9 8 7 6 5 4 3 2 1

Contents

Preface

An introductory course in ordinary differential equations has long been one of the staple courses in the undergraduate curriculum of mathematicians, engineers, and scientists. However, the contents of the standard course have changed in recent years. Today's course in differential equations incorporates more concepts from linear algebra, acknowledges the importance of nonlinear phenomena and reflects the increasing role of the computer in the study of mathematics.

This book recognizes these changes. Hence it may be used as a supplement to any of the current texts for a modern introduction to ordinary differential equations or it can be used as the primary source for such a course. Because of the many problems that are solved in detail, this book would be particularly useful for a program of self study. Alternatively it may be a convenient reference for those needing to recall or to quickly learn parts of the subject material. The focus is neither particularly theoretical nor completely applied but instead tries to strike a balance between these two points of view.

For example, chapter 2 and chapter 4 consider questions of a theoretical nature for first order and higher order differential equations respectively. In addition, solution techniques for several kinds of problems are explained. Chapters 3 and 5 then present numerous applications of first order and higher order equations. The solved examples in these chapters have been selected to illustrate how ordinary differential equations arise in mathematical models for physical systems and to show various ways in which the applied mathematician can extract information about the behavior of the physical system from the solution of the differential equation.

Chapter 6 develops the elementary theory for the Laplace transform and shows how this approach is particularly useful in exposing the relationships between input and output for linear physical systems.

Chapter 7 makes extensive use of linear algebra in presenting the theory and applications of the first order systems of linear ordinary differential equations, while chapter 8 sketches a brief introduction to nonlinear systems.

Numerical methods for approximating the solutions for linear and non-linear differential equations are introduced in chapter 10. The methods discussed are suitable for computer implementation and short BASIC programs are provided for some of the methods.

In addition to the solved examples found in each chapter, a number of supplementary problems with answers have been provided, where practical, in order that the students may test their grasp of the material.

For bearing with me during the writing of this book, I would like to thank my family, particularly Danielle for keeping me company through many long hours at the keyboard. In addition, I would like to thank Bill Hamill for his help through all the stages of development of the book .

1

Notation and Terminology

***I**n this chapter we introduce the terminology and the notation that we will be using throughout this text. In addition we describe the types of problems we will be solving and explain what we mean by a solution to a problem.*

TERMINOLOGY AND NOTATION

DIFFERENTIAL EQUATION

A *differential equation* is a single equation involving an unknown function and one or more of its derivatives. If the unknown function is a function of a single independent variable, then the derivatives are called *ordinary derivatives* and the equation is referred to as an *ordinary differential equation.* If the unknown function depends on more than a single variable, then the derivatives involved are *partial derivatives*, and the equation is called a *partial differential equation.* In this text, we shall confine our attention to ordinary differential equations.

If the unknown function is denoted by $y=y(x)$, then the derivatives of y will be denoted

$$\frac{dy}{dx} = y'(x),\ \frac{d^2y}{dx^2} = y''(x),\ \dots,\ \frac{d^Ny}{dx^N} = y^{(N)}(x).$$

SYSTEMS OF ORDINARY DIFFERENTIAL EQUATIONS

A *system of ordinary differential equations* is a set of N simultaneous ordinary differential equations, N>1, involving N unknown functions. We shall consider systems of equations in Chapter 7.

ORDER

The *order* of a differential equation is determined by the order of the highest order derivative that appears in the equation. If an equation contains a derivative of order N and no derivatives of order higher than N, we say the equation is an *equation of order N*.

LINEAR DIFFERENTIAL EQUATION

Consider the following equation of order N for the unknown function y=y(x):

$$F(x,y,y',\ldots,y^{(N)}) = 0 \tag{1.1}$$

If the function F is linear in y and all its derivatives, then we say the equation is a *linear differential equation*. The most general linear equation of order N is of the form

$$a_N\, y^{(N)}(x) + a_{N-1}y^{(N-1)}(x) + \ldots + a_1\, y'(x) + a_0\, y(x) = f(x). \tag{1.2}$$

Here, $a_k = a_k(x)$, k=0,1,...,N, in (1.2) are called the *coefficients* in the equation. In general, they are functions of the independent variable (only). In the special case that each of the coefficients is a constant, we refer to the equation as a *constant coefficient* differential equation. As we shall see, the solution of the equation is much easier in this special case. The term f(x) in (1.2) is called the *forcing term* in the equation. If f(x) is the zero function, then the equation is said to be *homogeneous*, and if f(x) is different from zero, then the equation is an *inhomogeneous*, differential equation.

NONLINEAR DIFFERENTIAL EQUATIONS

If the differential equation (1.1) includes any terms in which the unknown function y is multiplied by itself or any of its derivatives or if y or any of its derivatives appears as the argument in a nonlinear function, the equation is said to be *nonlinear*. In general, nonlinear problems are much more difficult to solve than corresponding linear problems. For that reason, we will devote the majority of our efforts to considering linear problems.

Example 1.1 Linear and nonlinear equations

1.1(A) LINEAR EQUATIONS

Each of the following is a linear equation for the unknown function y=y(x):

i) $y''(x) - xy(x) = \text{Sin } x$

ii) $(1+x^2)y'(x) - y(x) = 0$

iii) $2y^{(4)}(x) - 4y''(x) + 10y(x) = 1$

Equations i) and ii) have variable coefficients, while iii) is a constant coefficient equation. Equation ii) is homogeneous, the other two are inhomogeneous.

1.1(B) NONLINEAR EQUATIONS

Each of the following is a nonlinear equation for the unknown function $y=y(x)$:

i) $2yy' + y^2 = 0$

ii) $[1+(y')^2]^{1/2} = y(x)$

iii) $y''(x) + y'(x) = e^y$

Equation i) contains the nonlinear terms yy'' and y^2. In ii) and iii) the unknown function y or its derivative appear as the arguments of a nonlinear function.

WELL POSED PROBLEMS

In upcoming chapters, we shall consider various sorts of problems involving differential equations. The simplest sort of problem we shall consider is that of finding an unknown function $y=y(x)$ that satisfies a given differential equation. A function $y=y(x)$ is said to be a *solution* of the differential equation if it reduces the equation to an identity in x when $y(x)$ is substituted into the equation.

Example 1.2 Solutions to Equations

1.2(A) EQUATION OF ORDER ONE

For any choice of the constant C, the function $y = Ce^{3x}$ is a solution of the equation

$$y'(x) = 3y(x).$$

i.e., $$y'(x) = 3Ce^{3x} = 3y(x).$$

Hence $y'(x)$ is identically equal to $y(x)$ when $y(x)$ is given by Ce^{3x}, independent of the choice of the constant C.

1.2(B) EQUATION OF ORDER TWO

For any choice of the constants A and B, each of the functions

$$y_1 = A \text{ Sin } 2x \text{ and } y_2 = B \text{ Cos } 2x$$

is a solution for the equation

$$y''(x) + 4y(x) = 0.$$

i.e. $\quad y_1''(x) = -4A \text{ Sin } 2x = -4y_1(x),$

and $\quad y_2''(x) = -4B \text{ Cos } 2x = -4y_2(x)\,.$

Thus $y''(x) + y(x)$ vanishes identically for all values of x when

$$y(x) = A \text{ Sin } 2x \text{ or } y(x) = B \text{ Cos } 2x.$$

1.2(C) EQUATION OF ORDER FOUR

For any choice of the constants A,B,C,D each of the following functions

$$y_1(x) = A \text{ Sin } 2x \text{ and } y_2(x) = B \text{ Cos } 2x$$

$$y_3(x) = C \text{ Sinh } 2x \text{ and } y_4(x) = D \text{ Cosh } 2x,$$

satisfies the equation,

$$y^{(4)}(x) - 16\,y(x) = 0.$$

i.e. $\quad y^{(4)}(x) = 2^4\,y(x)\,.$

Recall that the hyperbolic sine and cosine are defined by

$$\text{Cosh } x = (e^x + e^{-x})/2$$

$$\text{Sinh } x = (e^x - e^{-x})/2\,.$$

We shall also be interested in problems in which the unknown function is required to satisfy certain *auxiliary conditions* in addition to satisfying the differential equation. Most problems arising in connection with applications involve auxiliary conditions.

Example 1.3 Auxiliary Conditions

1.3(A) ONE INITIAL CONDITION

The function $y(t) = 4e^{2t}$ solves the differential equation

$$y'(t) = 2y(t), \text{ for } t > 0,$$

and the *initial condition,*

$$y(0) = 4.$$

1.3(B) TWO INITIAL CONDITIONS

The function $y_1(x) = 5\text{Cos } 2x$ solves the differential equation

$$y''(x) + 4\,y(x) = 0, \text{ for } x > 0,$$

and the *initial conditions,*

$$y(0) = 5 \text{ and } y'(0) = 0.$$

Similarly, the function $y_2(x) = 5\text{Sin } 2x$ solves the differential equation

$$y''(x) + 4\,y(x) = 0, \text{ for } x > 0,$$

and the *initial conditions,*

$$y(0) = 0 \text{ and } y'(0) = 10.$$

1.3(C) BOUNDARY CONDITIONS

The function $y_3(x) = 5\text{Cos } 2x$ solves the differential equation

$$y''(x) + 4\,y(x) = 0, \text{ for } 0 < x < \pi,$$

and the *boundary conditions,*

$$y'(0) = 0 \text{ and } y'(\pi) = 0.$$

The auxiliary conditions in examples 1.3(a),(b) and (c) are referred to as *one point conditions* or, more commonly, as *initial conditions* because the conditions are all specified at a single point. Presumably this point can be interpreted as the "initial point" in the interval under consideration. In problems where auxiliary conditions are imposed, we are usually interested only in a specific range of values for the independent variable. The auxiliary conditions then specify the values of the solution or its derivatives (or combinations thereof) at one or both ends of this interval. In example 1.3(c) the independent variable is restricted to the interval $0 \le x \le \pi$, and the auxiliary conditions specify the values for the derivative of the solution at the endpoints (i.e. on the boundary) of this interval. For this reason, we refer to the auxiliary conditions in example 1.3(c) as *boundary conditions* or, less commonly as *two point conditions.*

INITIAL VALUE AND BOUNDARY VALUE PROBLEMS

A problem in which we are to find a function which satisfies one or more initial conditions in addition to a differential equation is referred to as an *initial value problem.* Similarly, a problem in which we are to find an unknown function that satisfies certain boundary conditions in addition to a differential equation, is called a *boundary value problem.*

GENERAL AND PARTICULAR SOLUTIONS

Note that the solutions in example 1.2 satisfy the differential equation for all values of the constants A,B, etc. In a sense then these are not single solutions but, in fact, families of solutions. Solutions containing arbitrary parameters, as in the case of the solutions in example 1.2, are referred to as *general solutions*. On the other hand, the solutions presented in example 1.3 contain no arbitrary constants. To be more precise, it is only for particular values of the arbitrary parameters that a general solution of the differential equation will satisfy the auxiliary conditions. For this reason, solutions to the differential equation which are subjected to auxiliary conditions are referred to as *particular solutions*. We will give more precise definitions of these terms in later chapters, but these definitions will serve for the present.

In the chapters to come we shall present various methods for constructing general and particular solutions to differential equations. If the problem involves no auxiliary conditions, then there may be more than a single solution; there will generally be an entire family of solutions. If the solution is subject to auxiliary conditions, then some of the solutions in the family of functions that solve the differential equation may be eliminated. If too many auxiliary conditions are imposed, *all* of the solutions may be eliminated so that the problem has *no* solution.

A problem consisting of a differential equation together with auxiliary conditions is said to be *well posed* if the number of auxiliary conditions is such that the problem has precisely one solution. That is, the conditions must not be so many that all solutions are precluded (i.e. a solution must *exist*) nor can the conditions be too few so that more than one solution is possible (i.e. the solution must be *unique*). We shall see that for linear problems, the problem is well posed if the number of auxiliary conditions is equal to the order of the equation.

Example 1.4 Well Posed Problems

Consider the following problems:

1.4(A)

Find $U=U(x)$ such that

$$U'(x) = U(x) \text{ for } x > 0$$

For any choice of the constant C, $U(x) = Ce^x$ solves the differential equation. There are infinitely many solutions $U(x)$ to this problem. This problem is not well posed because the solution is not unique. The number of auxiliary conditions is too few.

1.4(B)

Find V=V(x) such that

$$V'(x) = V(x), \text{ for } x > 0,$$

and $$V(0) = 2$$

For any choice of the constant C, $V(x) = Ce^x$ solves the differential equation, but only when C=2 is it the case that V(0)=2. This problem is well posed because it has one and only one solution.

1.4(C)

Find W=W(x) such that

$$W'(x) = W(x), \text{ for } x > 0,$$

and $$W(0) = 2, W'(0) = 3$$

Here $W(x) = Ce^x$ solves the differential equation for any choice of C. In addition, for C=2 we have W(0)=2. However, W′(0) does not equal 3 in this case. It is easy to see that there is no choice of C for which W(0)=2 and W′(0)=3 hence no solution exists for this problem. The number of auxiliary conditions is too many so the problem is not well posed.

NUMERICAL AND EXACT SOLUTIONS

Chapters 2 through 9 present a variety of methods for constructing solutions to differential equations and to initial value problems for differential equations. However, the solutions may arise in a form that is inconvenient for computational purposes; in addition there are problems to which none of these methods apply.

In such cases it may be desirable to use numerical means to construct a solution to a discrete version of the problem. That is, we replace the differential equation by an associated discrete equation whose solution involves only algebraic operations. Usually the computation is then carried out on a computer.

We refer to the solution of the problem in which the differential equation is actually solved as the *exact solution*. The solution of the corresponding discrete problem is referred to as the *numerical solution*. Constructing numerical solutions is the topic of chapter 10.

Supplementary Problems

1. By substituting into the differential equation, verify that for all choices of the constants A and B,

$$y(x) = A \text{ Sinh } x + B \text{ Cosh } x$$

solves

$$y''(x) - y(x) = 0 \text{ for } x > 0.$$

2. Which of the following problems for the unknown functions U, V, and W is well posed and which is not well posed,

(a) $U''(x) + U(x) = 0$, for $x > 0$

$U(0) = 1$

(b) $V''(x) + V(x) = 0$ for $x > 0$,

$V(0) = 1$, $V'(0) = 0$

(c) $W''(x) + W(x) = 0$

$W(0) = 1$, $W'(0) = 0$, $W''(0) = 2$.

2

First Order Equations

In this chapter we consider equations involving only the first derivative of the unknown function. We shall begin by considering linear equations, devising methods for explicitly constructing general and particular solutions for linear differential equations of order one. For nonlinear problems no general *method for constructing solutions can be found. Instead we consider a few special classes of nonlinear problems for which solution methods are available. We shall see that the solutions of nonlinear problems exhibit a wide range of complicated behavior not found in the solutions to linear problems.*

LINEAR EQUATIONS

The most general linear equation of order one may be written as follows,

$$a_1(x)y'(x) + a_0(x)y(x) = f(x). \tag{2.1}$$

Let A(x) denote an antiderivative of the function $-a_0(x)/a_1(x)$; i.e., let

$$A(x) = -\int a_0(x)/a_1(x)\,. \tag{2.2}$$

SINGULAR POINT

Note that A(x) is defined wherever $a_1(x)$ is different from zero. Any point at which $a_1(x)$ vanishes is said to be a *singular point* for the equation (2.1).

Also let F(x) denote an antiderivative of the "modified" forcing function, $e^{-A(x)}f(x)$; i.e. let,

$$F(x) = \int e^{-A(x)}\, f(x)/a_1(x)\,; \tag{2.3}$$

Theorem 2.1 General Solutions

Theorem 2.1 Suppose $a_0(x)$, $a_1(x)$ and f(x) are continuous functions. Then at any point at which $a_1(x)$ is not equal to zero, the *general solution* of equation (2.1) is the differentiable function given by

$$y(x) = C\, e^{A(x)} + e^{A(x)}\, F(x). \tag{2.4}$$

i.e. (2.4) solves equation (2.1) for all choices of the constant C, and every choice of the antiderivative F(x).

COROLLARY 2.1—PARTICULAR SOLUTIONS

Corollary 2.1 The unique *particular solution* of (2.1), which satisfies the auxiliary condition

$$y(x_0) = y_0 \tag{2.5}$$

is given by

$$y(x) = y_0 \exp[A(x)-A(x_0)] + e^{A(x)}\, F(x;x_0), \tag{2.6}$$

where $F(x;x_0)$ denotes the unique antiderivative of $e^{-A(x)}f(x)$, which vanishes at $x=x_0$; i.e.,

$$F(x;x_0) = \int_{x_0}^{x} e^{-A(z)}\, f(z)/a_1(z)\, dz \tag{2.7}$$

Remark: Theorem 2.1 states that (2.1) has an infinite family of solutions, all of the form given by (2.4). The corollary states further that the unique solution of the initial value problem consisting of (2.1) together with condition (2.5) is the solution given by (2.6). Problems 2.1 through 2.4, below, explain the development of solutions (2.4) and (2.6).

Note also that the theorem state that if $a_0(x)$, $a_1(x)$ and f(x) are continuous, then the solution to the differential equation can be found. It does not say that these functions *must* be continuous. Problem 2.5 in the solved problems illustrates how the lack of continuity in these functions affects the behavior of the solution.

Example 2.1 A First Order Problem

To find the general solution for the first order equation

$$y'(x) - 3y(x) = 1,$$

note first that A(x) = 3x and,

$$F(x) = \int e^{-3x} = -e^{-3x}/3.$$

Then, according to (2.4), the general solution of the equation is given by

$$y(x) = Ce^{3x} - 1/3.$$

It is easily checked that y(x) satisfies the differential equation for all choices of the arbitrary constant C.

To find the unique particular solution which satisfies the initial condition

$$y(0) = 10,$$

compute

$$F(x;0) = \int_0^x e^{-3z}\,dz = -(e^{-3x}-1)/3.$$

Then, according to (2.6), the particular solution is given by

$$\begin{aligned} y(x) &= 10\,e^{3x} - e^{3x}[e^{-3x}-1]/3 \\ &= 10\,e^{3x} - [1-e^{3x}]/3 = 31\,e^{3x}/3 - 1/3. \end{aligned}$$

Choosing C=31/3 in the general solution produces the particular solution that satisfies the initial condition. Note that instead of evaluating F(x;0), we could have used the general solution to write,

$$y(0) = C - 1/3 = 10.$$

Solving this equation for C leads to the unique choice of C for which the initial condition is satisfied.

NONLINEAR PROBLEMS

Theorem 2.1 states that equation (2.4) is the general solution for a general linear differential equation of order one. By varying the arbitrary constant C in (2.4), it is possible to obtain *all* solutions of (2.1).

No such general method exists for finding the solutions for first order equations that are nonlinear. Instead we shall discuss selected classes of nonlinear problems for which solution methods can be found. However, we do not speak of general solutions in connection with nonlinear problems, even for these special classes for which solution methods exist. The reason for this is that even if one is able to construct a solution containing an arbitrary constant, there may exist yet another solution that corresponds to no value of the arbitrary constant.

Separable Equations

A first order differential equation that can be written in the form

$$y'(x) = f(x)g(y). \tag{2.8}$$

is said to be *separable*. Since $y'(x)\,dx = dy$, it follows that

$$\int \frac{dy}{g(y)} = \int f(x)dx\,;$$

i.e.,

$$G[y(x)] = F(x), \tag{2.9}$$

where $G' = 1/g$ and $F' = f$. Note that (2.9) provides the general solution for (2.8) in *implicit form*; i.e. $y(x)$ is not expressed explicitly as a function of x, nor is there any guarantee that the antiderivatives F and G can be expressed in terms of a finite combination of elementary functions. Sometimes a solution of the form (2.9) is referred to as a *first integral* for the differential equation (2.8).

AUTONOMOUS EQUATIONS

A special subclass of the separable equations are the *autonomous* equations. These are equations of the form,

$$y'(x) = g(y)\,; \tag{2.10}$$

i.e. the independent variable x, does not explicitly appear in the equation. Autonomous equations have the advantage that it is often possible to predict the qualitative behavior of the solution without actually constructing the solution.

A point $y = Y_0$ is called a *critical point* for the equation (2.10) if $g(Y_0) = 0$. If $g(Y_0)=0$ then the constant function, $y(x)=Y_0$ for all x, is a solution of the differential equation. We refer to a constant solution of this sort as an *equilibrium solution*. It then is of interest to determine whether the equilibrium solutions are *stable* or *unstable*; i.e. do solutions whose initial value is near Y_0 tend toward Y_0 or away from Y_0 with increasing x. A detailed study of autonomous problems is the subject of Chapter 8.

Exact Equations

Suppose $y=y(x)$ can be expressed implicitly as a function of x by writing

$$F(x,y(x)) = 0$$

for some smooth function F of two variables. Then

$$dF/dx = F_x + F_y y'(x) = 0$$

and

$$y'(x) = -F_x(x,y)/F_y(x,y).$$

Here we are using the notation F_x and F_y to denote the first partials of F with respect to x and y respectively.

Now consider a differential equation of the form

$$y'(x) = -P(x,y)/Q(x,y) \tag{2.11}$$

for given functions P=P(x,y) and Q=Q(x,y). Based on our observations, it seems reasonable to expect this equation to have a solution, expressed implicitly by

$$F(x,y(x)) = 0 \tag{2.12}$$

if a smooth function F=F(x,y) exists such that $P=F_x$ and $Q=F_y$. If F is smooth then it follows that $F_{xy} = F_{yx}$. Hence in order to have $P=F_x$ and $Q=F_y$, it is necessary that P,Q satisfy

$$P_y = Q_x\,. \tag{2.13}$$

If condition (2.13) is satisfied, then it does, in fact, follow that the equation (2.11) has an implicit solution (2.12) where the function F can be determined from P and Q. An equation (2.11) for which (2.13) holds is said to be an *exact* equation.

Change of Variable

Separable equations and equations that are exact are two classes of nonlinear equations for which a systematic solution technique exists. Equations that are neither separable nor exact can sometimes be transformed into equations that are separable, exact, or even linear by a change of dependent variable.

For example, suppose the function f in the nonlinear equation

$$y'(x) = f(x,y) \tag{2.14}$$

is such that f is not a function of x and y independently. but instead, for nonzero constants A and B, f(x,y) = F(Ax+By) for some function F of one variable. Then the change of variable

$$u = Ax+By$$

implies

$$y'(x) = (u'(x) - A)/B$$

so that (2.14) reduces to

$$u'(x) = A + B\,F(u).$$

This is a separable equation for the unknown function u=u(x). If this equation can be integrated and u(x) found, then

$$y(x) = (u(x)-Ax)/B.$$

A similar substitution may be applied if $f(x,y) = F(y/x)$ for some single variable function F. In this case, we introduce the new unkown function $u(x) = y/x$, for which we have the identity, $y'(x) = u(x) + xu'(x)$. Then the change of dependent variable $u(x)=y/x$ reduces the nonlinear equation (2.14) to

$$u + x\,u'(x) = F(u);$$

i.e.,

$$\frac{u'(x)}{F(u) - u} = \frac{1}{x}.$$

Thus the transformation of dependent variable reduces the original equation for $y(x)$ to a separable equation for $u(x)$. If we are able to solve the equation for $u=u(x)$, then $y(x) = x\,u(x)$. Of course, there is no guarantee that we shall be able to integrate the transformed separable equation.

Note that $f(x,y)=F(y/x)$ for F a function of a single variable, if and only if

$$f(x,y) = f(kx,ky) \text{ for all nonzero } k. \tag{2.15}$$

A function f that satisfies (2.15) is said to be *homogeneous of degree zero* and the differential equation is said to be a homogeneous nonlinear equation. (The term homogeneous has more than one meaning in connection with differential equations.)

SOLVED PROBLEMS

Solution of first order linear equations

PROBLEM 2.1

Find the general solution of first order homogeneous equation

$$a_1(x)y'(x) + a_0(x)y(x) = 0. \tag{1}$$

Also, find the particular solution that satisfies the auxiliary condition (2.5); i.e., $y(x_0) = y_0$.

SOLUTION 2.1

Let

$$A(x) = -\int a_0(x)/a_1(x). \tag{2}$$

At each point x that is not a singular point, (1) is equivalent to

$$\frac{y'(x)}{y(x)} = -\frac{a_0(x)}{a_1(x)}. \tag{3}$$

Since $\ln[y(x)]$ is an antiderivative of the left side of (3) and $A(x)$ is an antiderivative of the right side, it follows that at each nonsingular point x,

$$\ln y(x) = A(x) + \text{constant}. \tag{4}$$

Since the constant here is arbitrary, (4) is equivalent to

$$y(x) = Ce^{A(x)}. \tag{5}$$

where $C = e^{\text{constant}}$.

The constant C in (5) is arbitrary, hence (5) represents an infinite family of solutions as C is allowed to range over all possible values. Substituting x_0 into (5) and using the intial condition to solve for C, we find that for the following particular choice of C,

$$C = y_0/\exp[A(x_0)], \tag{6}$$

the solution (5) satisfies the auxiliary condition (2.5).

We have shown that if (1) has any solution, then this solution must be given by (5). In addition, the choice (6) for C is the one and only choice for which a solution of the form (5) can satisfy the auxiliary condition (2.5).

PROBLEM 2.2

Solve the initial value problem for the unknown function $y=y(x)$,

$$(1-x^2)y'(x) - 2y(x) = 0, \text{ for } x > 0, \; y(0) = 7.$$

SOLUTION 2.2

Note that the points $x = 1,-1$ are singular points for the equation. Only the point $x=1$ lies in the interval on which the solution is to be found. At nonsingular points, we find the general solution to the equation to be given by

$$y(x) = C\,\frac{1+x}{1-x},$$

and $C=7$ is the unique choice of C for which the initial condition is satisfied.

This example suggests that at a singular point it may happen that the general solution of the linear differential equation becomes undefined. Then it is generally not appropriate to impose auxiliary conditions at singular points of the equation. We can characterize the behavior of the solution near a singular point more precisely using the power series methods of Chapter 6.

PROBLEM 2.3

Find the general solution of the inhomogeneous equation (2.1). Also, find the particular solution that satisfies the auxiliary condition (2.5).

SOLUTION 2.3

In the homogeneous case that f(x) is identically zero, the solution of (2.1) has been found to be,

$$y(x) = C\, e^{A(x)}.$$

Then, when f(x) is not the zero function, we are motivated to try to find a function C(x) such that

$$y(x) = C(x)\, e^{A(x)} \qquad (1)$$

solves (2.1). This technique of seeking a solution for the inhomogeneous equation that is a variation of the homogeneous solution is known as *variation of parameters*. We shall see this technique again in connection with inhomogeneous equations of higher order.

If y(x) is given by (1), then

$$\begin{aligned} y'(x) &= C'(x)e^{A(x)} + C(x)e^{A(x)}A'(x) \\ &= C'(x)e^{A(x)} - C(x)e^{A(x)}a_0(x)/a_1(x) \\ &= C'(x)e^{A(x)} - a_0(x)y(x)/a_1(x). \end{aligned}$$

Hence

$$a_1(x)y'(x) + a_0(x)y(x) = a_1(x)e^{A(x)}C'(x).$$

But, it follows from (2.1) that

$$a_1(x)e^{A(x)}C'(x) = f(x);$$

i.e.,

$$C'(x) = e^{-A(x)}f(x)/a_1(x)$$

Taking antiderivatives on each side of this equation leads to,

$$x\, C(x) = \int^{x} e^{-A(z)}f(z)/a_1(z)\, dz\,. \qquad (2)$$

Using (2) in (1) gives us a solution for (2.1). It is customary, however, to include the general homogeneous solution with the particular solution (1) and to give (2.4) as the general solution for (2.1). We have shown that if (2.1) has any solution, then this solution is of the form (2.4) for some choice of C and antiderivative F(x).

In order to satisfy the auxiliary condition (2.5), we recall that the homogeneous solution,

$$y_H(x) = y_0 \exp[A(x) - A(x_0)] \tag{3}$$

satisfies (2.5). If we choose the antiderivative (2) so that

$$C(x) = \int_{x_0}^{x} e^{-A(z)} f(z)/a_1(z)\, dz, \tag{4}$$

then $C(x_0)=0$; if we write

$$y_P(x) = C(x)e^{A(x)}, \tag{5}$$

then

$$y(x) = y_H(x) + y_P(x) \tag{6}$$

is the unique function that satisfies (2.1) and (2.5).

We may view $y_H(x)$ given by (3) as the part of the solution reflecting the response to the initial condition (2.5) and $y_P(x)$ as representing the response to the forcing term $f(x)$.

PROBLEM 2.4

Solve the following inhomogeneous initial value problem for the unknown function $y=y(x)$:

$$y'(x) - 4y(x) = \text{Sin } 2x, \text{ for } x > 0,$$

$$y(0) = 5.$$

SOLUTION 2.4

We solve first for $y_H(x)$, the solution to the homogeneous equation; i.e.,

$$y_H'(x)/y_H(x) = 4$$

implies

$$\ln |y_H(x)| = 4x + c.$$

Then

$$y_H(x) = Ce^{4x}.$$

At $x=0$ this becomes $y_H(0) = C$. Comparing this with the initial condition $y_H(0) = 5$ leads to the conclusion that $C=5$.

Next, substituting $y_P(x) = C(x)e^{4x}$ into the inhomogeneous differential equation leads to

$$C'(x) = e^{-4x} \text{ Sin } 2x;$$

i.e.,

$$C(x) = \int_0^x e^{-4z} \text{ Sin } 2z\, dz$$

$$= 1/10 - e^{-4x}(4 \text{ Sin } 2x + 2 \text{ Cos } 2x)/20$$

Then

$$y_P(x) = 1/10\, e^{4x} - (4 \text{ Sin } 2x + 2 \text{ Cos } 2x)/20$$

and

$$y(x) = 5.1\, e^{4x} - .05\,(4 \text{ Sin } 2x + 2 \text{ Cos } 2x).$$

PROBLEM 2.5

Find the general solution of the homogeneous linear equation

$$a_1(x)y'(x) + a_0(x)y(x) = 0, \text{ for } -\infty < x < \infty,$$

in the following cases:

(a) $a_0(x) = x$ $\qquad a_1(x) = 1$

(b) $a_0(x) = \begin{cases} 0 & \text{if } x < 0 \\ x & \text{if } x \geq 0 \end{cases}$ $\qquad a_1(x) = 1$

(c) $a_0(x) = \begin{cases} 0 & \text{if } x < 0 \\ 1-x & \text{if } x \geq 0 \end{cases}$ $\qquad a_1(x) = 1$

(d) $a_0(x) = 1$ $\qquad a_1(x) = x$

Note: In this problem, we not only solve several homogeneous equations, we examine the effect on the solution of allowing the coefficient $a_0(x)$ to have varying degrees of irregular behavior. In (a) $a_0(x)$ is everywhere continuous; in (b) $a_0(x)$ is continuous but $a_0'(x)$ is discontinuous at x=0. In (c) $a_0(x)$ is discontinuous at x=0, and in (d), x=0 is a singular point.

SOLUTION 2.5

The solution is given in four parts.

part (a) We rewrite the equation as

$$y'(x)/y(x) = -x$$

Then,

$$\ln y(x) = -x^2/2 + c$$

and

$$y(x) = C \exp[-x^2/2].$$

part (b) We have,

$$y'(x)/y(x) = \begin{cases} 0 & \text{if } x < 0 \\ -x & \text{if } x > 0 \end{cases}$$

Hence,

$$\ln|y(x)| = \begin{cases} c_1 & \text{if } x < 0 \\ -x^2/2 + c_2 & \text{if } x > 0 \end{cases}$$

This is equivalent to

$$y(x) = \begin{cases} C_1 & \text{if } x < 0 \\ C_2 \exp[[-x^2/2] & \text{if } x > 0 \end{cases}$$

This solution is continuous at x=0 if we choose $C_1 = C_2$. If we make this choice then it automatically follows that $y'(x)$ is continuous at x=0. If we choose C_1 not equal to C_2 then y(x) satisfies the equation for x<0 and for x>0 but at x=0, y(x) is not continuous and $y'(x)$ is not defined. Hence, to obtain a solution for all x, $-\infty<x<\infty$, we must choose $C_1 = C_2$.

part (c) We have,

$$y'(x)/y(x) = \begin{cases} 0 & \text{if } x < 0 \\ x-1 & \text{if } x > 0 \end{cases}$$

Therefore,

$$\ln|y(x)| = \begin{cases} c_1 & \text{if } x < 0 \\ x^2/2 - x + c_2 & \text{if } x > 0. \end{cases}$$

This is equivalent to

$$y(x) = \begin{cases} C_1 & \text{if } x < 0 \\ C_2 \exp(-x+x^2/2) & \text{if } x > 0 \end{cases}$$

$C_1=C_2$ is the only choice of C_1 and C_2 for which y(x) is continuous at x=0 and you will note that $y'(x)$ is not continuous at x=0 for this choice.

part (d) Note that x=0 is a singular point; i.e. $a_1(x)=x$ and $a_0(x)=1$ is equivalent to $a_1(x)=1$ and $a_0(x)=1/x$. Then for x not equal to zero, we have

$$y'(x)/y(x) = -1/x$$

It follows that

$$\ln|y(x)| = -\ln|x| + c.$$

For x not equal to zero, the solution is given by

$$y(x) = C/x.$$

In summary, note that in (a) where $a_0(x)$ a smooth function, the solution y(x) is also smooth. In (b), $a_0(x)$ is piecewise defined; i.e., $a_0(x)$ is continuous for all x, but $a_0'(x)$ is discontinuous at x=0. Then the solution is continuous with a continuous derivative if we choose $C_1=C_2$; however, y″(x) is not continuous at x=0. Next, in case (c) the coefficient $a_0(x)$ is discontinuous at x=0 and this causes y′(x) to be discontinuous at x=0, although the solution y(x) is continuous for all x. Finally, in case (d), x=0 is a singular point; i.e., $a_0(x)$ becomes undefined at x=0. Then the solution y(x) is singular (becomes undefined) at x=0 as well.

PROBLEM 2.6

Solve the initial value problem for the unknown function y=y(x):

$$y'(x) + Ay(x) = e^{-x}, \text{ for } x > 0,$$
$$y(0) = 1$$

with the coefficient $a_0(x)=A$ equal to the following constant values: A = 0,1,2.

SOLUTION 2.6

The solution of the homogeneous equation is easily found to be

$$y_H(x) = C\, e^{-Ax}.$$

If we suppose

$$y_P(x) = C(x)\, e^{-Ax},$$

then, on substituting into the inhomogeneous equation, we conclude that

$$C'(x) = \exp[(A-1)x].$$

For A not equal to 1, this implies

$$C(x) = \exp[(A-1)x]/(A-1)$$

and

$$y_P(x) = e^{-x}/(A-1).$$

When A=1, we have

$$C'(x) = 1$$

and

$$C(x) = x$$
$$y_P(x) = xe^{-x}.$$

Then the general solution to the inhomogeneous differential equation is as follows

$$y(x) = \begin{cases} C e^{-Ax} + e^{-x}/(A-1) & \text{for } A \text{ not equal to } 1 \\ C e^{-Ax} + x e^{-x} & \text{for } A = 1. \end{cases}$$

In particular, for A=0,1,2

$$y_0(x) = C - e^{-x}$$

$$y_1(x) = C e^{-x} + x e^{-x}$$

$$y_2(x) = C e^{-2x} + e^{-x}$$

The solution to the initial value problem for each of the three choices for A is then found by choosing C so that the initial condition is satisfied; i.e.,

$y_0(0) = C - 1 = 1$ implies C=2; so

$$y_0(x) = 2 - e^{-x}.$$

$y_1(0) = C - 0 = 1$ implies C=1; so

$$y_1(x) = e^{-x} + x e^{-x}$$

$y_2(0) = C + 1 = 1$ implies C=0; so

$$y_2(x) = e^{-x}.$$

Note that in the case A=1, the forcing function in the inhomogeneous equation is a solution for the homogeneous equation. This causes the solution $y_P(x)$ to have a different form than it does when the forcing term is not a solution for the homogeneous equation.

Solutions of nonlinear equations

PROBLEM 2.7

Consider the problem of finding y=y(x) such that

$$y'(x) = 3x^2y^2, \; y(0) = 1.$$

SOLUTION 2.7

We can separate the variables in this equation to obtain

$$\int y^{-2}\, dy = \int 3x^2\, dx\,;$$

i.e.,

$$-1/y = x^3 + c.$$

We can solve this equation for y in terms of x to obtain

$$y(x) = \frac{-1}{x^3 + c}$$

Choosing c = –1 causes the initial condition to be satisfied. Note that the solution of this initial value problem,

$$y(x) = 1/(1-x^3),$$

becomes undefined at x=1. This singular point occurs at the point where the coefficient of $y'(x)$ vanishes; in this nonlinear equation, that coefficient is $1/y^2$ so there is no way to know that the coefficient vanishes until after the solution is found.

Note further that if the initial condition is changed to $y(0) = 2$, then the solution of the initial value problem changes to

$$y(x) = 2/(1-2x^3),$$

which becomes undefined at $x=(1/2)^{1/3}$. This is an additional way in which nonlinear problems differ from linear problems. Not only can singular points arise unexpectedly, their location may depend on the initial condition.

PROBLEM 2.8

Solve the separable equations

$$y'(x) = \frac{5x^4 - 4x^3}{5y^4 + 1}$$

and

$$y'(x) = \exp(-y^2)(1 + \ln x + x^3)^{1/2}$$

SOLUTION 2.8

By separating variables and integrating, we see that the first equation has the implicit solution

$$y^5 + y = x^5 - x^4 + c.$$

There is no way we can solve for y in terms of x, nor can we solve explicitly for x in terms of y.

In the case of the second equation, we discover that we are unable to find antiderivatives in the separated equation,

$$\int \exp(y^2)\, dy = \int (1 + \ln x + x^3)^{1/2} dx\,.$$

These examples have been deliberately constructed to illustrate some of the things that can go wrong in solving separable equations.

PROBLEM 2.9

Solve the following initial value problem for the unknown function P(t):

$$P'(t) = kP(1-P),\ P(0) = P_0\,.$$

SOLUTION 2.9

The equation here is separable; i.e., we can write

$$\int \frac{dP}{P(1-P)} = k\int dt.$$

We can use partial fractions (see also chapter 6) to write

$$\frac{1}{P(1-P)} = [\,1/P + 1/(1-P)\,]$$

Hence,

$$\ln\,[P/(1-P)] = kt + c\,;$$

and

$$P/(1-P) = C\exp(kt).$$

Now we can solve explicitly for P in terms of t,

$$P(t) = \frac{C}{C + \exp(-kt)}\,.$$

Finally, as long as P_0 is not equal to 1, we can use the initial condition to solve for C:

$$C = P_0/(1-P_0),$$

which leads to the following solution for the initial value problem,

$$P(t) = \frac{P_0}{P_0 + (1-P_0)\exp(-kt)}\,.$$

Note that if $P_0=0$ or $P_0=1$ then $P(t) = P_0$ for $t\geq 0$ is an equilibrium solution of the initial value problem; i.e. $P=0$ and $P=1$ are critical points for $F(P) = kP(1-P)$.

PROBLEM 2.10

Solve the equation

$$y'(x) = \frac{x^3 - y}{y^3 + x}\,.$$

SOLUTION 2.10

Letting $P(x,y)= y-x^3$ and $Q(x,y)= x+y^3$, we can easily see that condition (2.13) is satisfied; i.e. this equation is exact.

To find $F=F(x,y)$, we write

$$F_x = P = y-x^3.$$

Integrating both sides of this equation with respect to x (while holding y fixed) leads to

$$F(x,y) = xy - x^4/4 + p(y),$$

where $p=p(y)$ denotes an arbitrary function of y. Now differentiating this expression with respect to y leads to

$$F_y = x + p'(y) = Q(x,y) = x + y^3; \text{ thus } p'(y) = y^3.$$

Then $p(y) = y^4/4 + C$ for arbitrary constant C, and it follows that

$$F(x,y) = xy + (y^4-x^4)/4 + C.$$

This is the implicit solution (2.12) for the differential equation. Note that we could have as well solved the problem by intgrating Q with respect to y (holding x fixed) to find

$$F(x,y) = xy + y^4/4 + q(x).$$

for q(x) an arbitrary function of x. Then, differentiating this with respect to x leads to

$$F_x = y + q'(x) = P = y - x^3.$$

It follows that $q(x) = -x^4/4 + C$, which produces the result we obtained before.

PROBLEM 2.11

Solve the equation,

$$y'(x) = (y^2+xy)/x^2 .$$

SOLUTION 2.11

Letting $P(x,y) = -(y^2+xy)$ and $Q(x,y) = x^2$, we find that (2.13) is not satisfied; i.e. this equation is not exact.

Note, however, that

$$\frac{y^2+xy}{x^2} = \frac{(y^2+xy)/(xy^2)}{x^2/(xy^2)} = \frac{1/x + 1/y}{x/y^2} .$$

and if we let $P(x,y) = -1/x - 1/y$ and $Q(x,y) = x/y^2$, then (2.13) is satisfied and the equation is exact. The factor $1/(xy^2)$, by which we multiplied the original P and Q in order to obtain a pair of functions satisfying (2.13), is called an *integrating factor*. Sometimes an equation of the form (2.8) that is not exact to begin with can be made exact by multiplying by an appropriate integrating factor.

Using $F_x = P = -1/x - 1/y$, and integrating with respect to x while holding y constant, leads to

$$F(x,y) = -\ln|x| - x/y + p(y).$$

We differentiate with respect to y and set the result equal to Q. This yields

$$x/y^2 + p'(y) = x/y^2, \text{ so}$$

$$p'(y) = 0 \text{ and } p(y) = \text{constant.}$$

Then

$$F(x,y) = -\ln|x| - x/y + C,$$

and the solution of the differential equation is given implicitly by

$$\ln|x| + x/y = C \text{ for } x,y \text{ not equal to zero.}$$

Note that $y(x) = 0$ is also a solution to the differential equation, but this solution is not obtained from the implicit solution above for any value of the constant C.

PROBLEM 2.12

Solve for $y=y(x)$ if

$$x\,y'(x) = y + [x^2+y^2]^{1/2}.$$

SOLUTION 2.12

Rewrite the equation as,

$$y'(x) = y/x + [1 + (y/x)^2]^{1/2}.$$

Then the equation is seen to be a homogeneous nonlinear equation and the change of variables $u = y/x$ reduces it to

$$x\,u'(x) + u = u + [1+u^2]^{1/2}$$

$$[1+u^2]^{-1/2}\,u'(x) = 1/x\,.$$

Integration leads to

$$\ln|u + [1+u^2]^{1/2}| = \ln|x| + c$$

$$u + [1+u^2]^{1/2} = Cx,$$

or

$$y + [x^2+y^2]^{1/2} = Cx^2.$$

PROBLEM 2.13

Use a change of variable to solve the differential equation

$$y'(x) = \text{Cos}(x+y)$$

SOLUTION 2.13

Letting $u(x)=x+y$ so that $u'(x)= 1+y'(x)$, we get

$$u'(x) = 1 + \text{Cos}\, u.$$

This equation is separable, and we find after integration,

$$\text{Sin}\, u/(1 + \text{Cos}\, u) = x + C; \text{ or}$$

$$\text{Sin}\,(x+y)/(1 + \text{Cos}(x+y)) = x + C.$$

There does not appear to be any way to obtain y explicitly in terms of x.

PROBLEM 2.14

Use a change of variable to solve the following initial value problem,

$$y'(x) = \frac{x-y}{x+2y},\ y(0) = 4.$$

SOLUTION 2.14

Applying the criterion (2.15), it is evident that this equation is a nonlinear homogeneous equation. Then let $u(x) = y/x$ so that the equation reduces to,

$$x\,u'(x) = \frac{1-2u-2u^2}{1+2u}.$$

This equation separates to

$$\int \frac{1+2u}{1-2u-2u^2}\,du = \int \frac{1}{x}\,dx,$$

and we integrate to obtain

$$\ln|1-2u-2u^2| = -2\ln|x| + c.$$

Then

$$1-2u-2u^2 = C/x^2,$$

or

$$x^2-2xy-2y^2 = C.$$

If $y(0)=4$, then

$$0 - 0 - 2(4)^2 = C,$$

and

$$2y^2 + 2xy - x^2 = 32\,.$$

This can be solved to obtain y explicitly in terms of x,

$$y = -x/2 \pm (3x^2+64)^{1/2}/2\,.$$

The first order linear equation (2.1) has a family of general solutions given by (2.4). There is a unique member of this family that satifies the initial condition (2.5). This particular solution is given by (2.6). These solutions are well behaved except possibly at singular points; i.e. points at which $a_1(x)=0$.

There is no general method for constructing solutions to nonlinear equations. If the nonlinear equation happens to be separable or exact then we can construct a first integral which provides at least an implicit solution. In addition, some nonlinear equations may be reduced to equations that are separable, exact or even linear by a change of variables. Except in a few simple cases there is no systematic way to anticipate when a change of variables will cause such a reduction.

Nonlinear equations may develop spontaneous singularities and the location of these singularities may depend on the initial conditions. It is not appropriate to speak of general solutions in connection with nonlinear equations. Even when a solution containing an arbitrary constant can be found for a nonlinear equation, it may be that there are additional solutions to the equation that do not correspond to any choice of the arbitrary constant.

SUPPLEMENTARY PROBLEMS

Find the general solution or the indicated particular solution for each of the following:

2.1	$x\,y'(x) - 4\,y(x) = x^5$	
2.2	$y'(x) - (\text{Sin } x)\,y(x) = \text{Sin } 2x$	
2.3	$x^2\,y'(x) - (x+1)\,y(x) = x$	$y(1) = -2$
2.4	$L\,i'(t) + R\,i(t) = E\,\text{Sin } \Omega t$	$i(0) = 0$
2.5	$(x+4y+2)\,y'(x) = (x+y-1)$	
2.6	$(x+y+1)^2 y'(x) = (x+y-1)^2$	
2.7	$y'(x) + y(x) = x$	
2.8	$y'(x) + y(x) = x^2+2$	
2.9	$y'(x) - y(x) = 0$	$y(0) = 1$
2.10	$x\,y'(x) + 5y(x) = 6x^2$	$y(1) = 3$
2.11	$y'(x) + y(x) = e^{-x}$	$y(0) = 1$
2.12	$y'(x) - 2y(x) = 1$	
2.13	$y'(x) - 2y(x) = x^2+x$	
2.14	$y'(x) + 2xy(x) = x$	
2.15	$x\,y'(x)+y(x) = 3x^3-1$	
2.16	$y'(x) + e^x y(x) = 3e^x$	
2.17	$(x^2+y^2)y'(x) = -2xy(x)$	
2.18	$(y-x)y'(x) + y(x) + x = 0$	
2.19	$x\,y'(x) = y + y^2$	
2.20	$y'(x) + 2y(x) = 3e^{-2x}$	

ANSWERS TO SUPPLEMENTARY PROBLEMS

2.1 $y(x) = Cx^4 + x^5$

2.2 $y(x) = C \exp[-\text{Cos } x] - 2\text{Cos } x + 2$

2.3 $y(x) = -x - x \exp[1-1/x]$

2.4 $i(t) = \{E(R \text{ Sin } \Omega t - \Omega L \text{ Cos } \Omega t) + E\Omega L \exp[-Rt/L]\}/(R^2+\Omega^2 L^2)$

2.5 $(x+2y)(x-2y-4)^3 = C$ (implicit solution)

2.6 $(x+y)^2 + 1 = C e^{x-y}$ (implicit solution)

2.7 $y(x) = C e^{-x} + x - 1$

2.8 $y(x) = C e^{-x} + (x - 2)^2$

2.9 $y(x) = e^x$

2.10 $y(x) = (22 x^{-5} + 6x^2)/7$

2.11 $y(x) = (x+1)e^{-x}$

2.12 $y(x) = Ce^{2x} - 1/2$

2.13 $y(x) = Ce^{2x} - (x^2+2x+1)/2$

2.14 $y(x) = C \exp[-x^2] + 1/2$

2.15 $y(x) = C x^{-1} + (3x^3-4)/4$

2.16 $y(x) = C \exp[-e^x] + 3$

2.17 $y^3 + 3x^2y = C$

2.18 $\text{ArcTan}(y/x) - \text{Log}(x^2+y^2) = C$

2.19 $y(x) = x(C-x)^{-1}$

2.20 $y(x) = Ce^{-2x} + 3x e^{-2x}$

3

Applications of First Order Equations

The previous chapter was devoted to developing methods for solving first order differential equations. We showed how to construct the solution for homogeneous and inhomogeneous linear equations and we found methods for solving a few specific types of nonlinear equations. In this chapter we show how the solution methods may be applied to obtain information of practical importance relating to various applications. These applications include:

GROWTH AND DECAY

radioactive decay

biological growth

HEAT TRANSFER—Newton's law of cooling

MIXING—rate of increase = input – output

MECHANICS—Newton's laws of motion

GEOMETRIC APPLICATIONS—orthogonal trajectories

The examples in this chapter are intended to illustrate how differential equations arise naturally in modelling physical systems and how useful information about the physical system can be extracted from the solution once it has been found.

GROWTH AND DECAY

Radioactive Decay

Certain atoms are inherently unstable; from time to time they spontaneously tranform to an atom of a different (lighter) element. The excess mass is emitted in the form of radiation. Such elements are called *radioactive*. The rate at which this transformation takes place is called the *decay rate*, and it has been experimentally observed that the decay rate is proportional to the amount of radioactive material present. If A(t) denotes the amount of mass of a given radioactive material that is present at time t, then the derivative of A(t) represents the decay rate of the material. For a radioactive material this derivative is negative since the amount of material present is decreasing with time. Then the empirical observation that the decay rate is proportional to the present amount can be expressed by,

$$dA/dt = -k\,A(t), \tag{3.1}$$

where k denotes the positive constant of proportionality. For elements that decay very rapidly, k is a large positive number and for elements that decay slowly, k is a small positive number.

Thus, A(t), the amount of radioactive material present at time t, is governed by a first order, homogeneous linear differential equation with constant coefficients. Then, according to Theorem 2.1, the solution is given by

$$A(t) = A(0)e^{-kt}, \tag{3.2}$$

where A(0) denotes the amount of material present at t=0.

HALF LIFE

The *half-life* of a radioactive element is defined to be the amount of time required for the amount of material to decay to one half of the original amount present. That is, the half-life is the time T at which $A(T) = \frac{1}{2}A(0)$. Since A(t) is given by (3.2), this implies that T must satisfy,

$$\tfrac{1}{2} = e^{-kT}, \text{ so}$$

$$kT = \ln 2. \tag{3.3}$$

From this last equation, we can compute the rate constant k if T is known, or we can compute T if k is known.

Population Models

Differential equations may be used to model the growth of populations of various sorts. We may consider populations of humans or populations of insects or other biological organisms. The mathematical description of a population's growth is not affected by the nature of the individuals comprising the population but only by the factors that influence the population's growth.

BIRTH AND DEATH RATE

In simplest terms, the population size changes in two ways: individuals die and others are born. The number of individuals that die in a year is proportional to the population size, and the constant of proportionality is the *death rate*, d. Similarly, the number of births in a year is proportional to the population size and that constant of proportionality is *the birth rate*, b. Since b and d are specified in units like "number of births or deaths per thousand individuals," we can write

$$P'(t) = b\,P(t) - d\,P(t)$$

where P(t) denotes the population size at time t. That is, the rate at which the population size is changing is equal to the difference between the birth rate and the death rate. If we let $k = b - d$, then P(t) satisfies the following linear, constant coefficient differential equation:

$$P'(t) = k\,P(t). \qquad (3.4)$$

This is the same equation that governs radioactive decay, but here the constant multiplier of the unknown function is positive if b is greater than d and negative if b is less than d; in the decay equation the multiplier, $-k$, is always negative. A positive multiplier is consistent with a steadily growing value of the unknown function, while a negative multiplier produces a solution that decreases steadily with time. Figure 3.1 shows population size P(t) plotted against time in each of the three cases: $b>d$, $b=d$ and $b<d$.

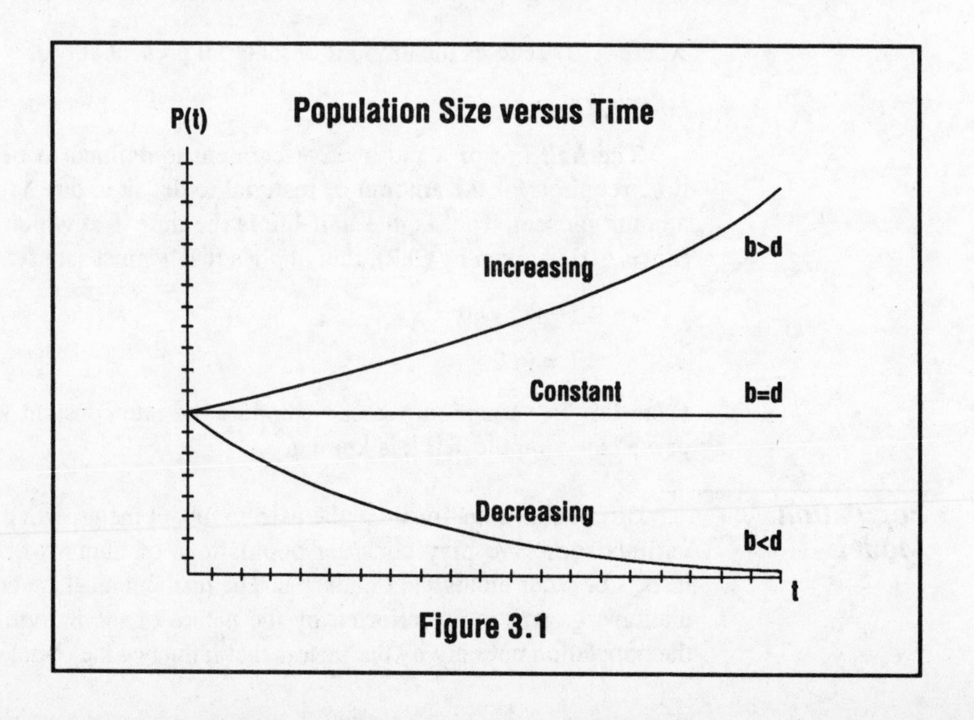

Figure 3.1

VARIABLE GROWTH RATE

We have supposed here that b and d (and hence k) are constants. Over short periods of time it may be an acceptable approximation of reality to suppose k is constant, but it is more realistic to assume that k varies with P. For example, the assumption that for positive constants K and P*, the growth rate is given by

$$k = K\,(P^* - P(t)) \tag{3.5}$$

implies that the growth rate k is positive (so the population is increasing) as long as P(t) does not exceed some "population limit" given by P*. If P(t) exceeds P*, then the growth rate becomes negative and the population decreases as long as P(t) continues to exceed P*. In particular, the population does not increase without bound nor does it decrease steadily to zero.

HEAT TRANSFER

A cup of hot coffee, if left to sit, eventually cools off until it reaches room temperature. Similarly, a glass of ice water, left to sit, will gradually warm up until it too has reached room temperature. Each of these processes proceeds according to Newton's law of cooling.

NEWTON'S LAW OF COOLING

Newton's law of cooling states that the rate at which the temperature of a body changes is proportional to the difference in temperature between the body and its surroundings. If we denote the temperature of the body by T(t) and the temperature of the region surrounding the body by T*, then Newton's law can be expressed,

$$T'(t) = -k\,[T(t) - T^*]. \tag{3.6}$$

Here k denotes a positive constant, so if the body is hotter than its surroundings (like the cup of coffee) then T(t)–T* is positive and T′(t) is negative; i.e., the temperature is decreasing. Similarly, if the body is cooler than its surroundings (like the glass of ice water) T(t)–T* is negative and T′(t) is positive; the temperature is increasing and the body is growing warmer. In either case, the instantaneous temperature T(t) satisfies a first order linear, inhomogeneous differential equation.

CHEMICAL MIXING

Consider a container filled with a mixture of chemical X. Suppose the container has an inlet pipe through which another chemical mixture is supplied at a fixed flow rate Q. The two mixtures are assumed to become immediately, uniformly mixed. At the same time the now uniformly mixed chemical is piped out of the container at a flow rate of Q so that the total volume of mixture in the container remains fixed. Then the amount of chemical X in the container can be shown to satisfy a first order differential equation.

Let X(t) denote the amount of chemical X in the container at time t, and let V denote the volume of the container. Then the concentration of chemical X at time t is,

$$C(t) = X(t)/V .$$

Then chemical X is flowing out of the container at the rate of

$$\text{outflow} = QC(t) = X(t)Q/V$$

units of mass per unit of time. If the chemical mixture flowing into the container contains chemical X at the concentration level f(t), for f(t) given, then chemical X is flowing into the container at the rate of

$$\text{inflow} = Qf(t)$$

units of mass per unit of time. Then the time rate of change of amount of chemical X in the container is equal to the difference between inflow and outflow; i.e.,

$$X'(t) = Q\,f(t) - Q/V\,X(t), \tag{3.7}$$

$$\text{rate of increase} = \text{inflow} - \text{outflow}.$$

Thus, X(t) solves a linear first order, inhomogeneous differential equation. Note that if inflow exceeds outflow then X′(t) is positive, consistent with increasing X(t).

MECHANICS

NEWTON'S SECOND LAW OF MOTION

Newton's second law of motion states that the rate of change of a body's momentum with respect to time is equal to the net force acting on the body.

The momentum of a body is defined as the product of its mass and its velocity. Thus, Newton's law states

$$d/dt\,(mv) = F. \tag{3.8}$$

For bodies with constant mass, this is just the statement that the rate of change of velocity is proportional to the net force acting on the body and that the constant of proportionality is the body's mass. Equation (3.8) is called the *equation of motion* for the system under consideration.

EQUATION OF MOTION: ALTERNATE FORM

The velocity v is equal to the rate of change of position with respect to time. For motion in a straight line this becomes

$$v = dx/dt,$$

and hence in cases in which the velocity depends explicitly on position, we have, according to the chain rule,

$$\frac{d}{dt}\,v[x(t)] = \frac{dv}{dx}\frac{dx}{dt} = \frac{dv}{dx}\,v\,.$$

Then for a body with constant mass, the equation of motion has the following alternate form

$$m\,v\,dv/dx = F. \tag{3.9}$$

The force F may depend explicitly on the time t, the velocity v, the position x, or a combination of these. In the event that F is an explicit function of t only, we use the form (3.8) of the equation of motion. If F is an explicit function of v only, then equation (3.9) is indicated. In either case, we are faced with a nonlinear first order equation in which the variables are separable. If F is an explicit function of x only, then we replace v in (3.8) by dx/dt. This leads to an equation of second order. These equations are discussed in chapters 4 and 5. In the event that F depends explicitly on more than one of the variables t, x, or v, we are faced with a problem beyond the scope of this text.

GEOMETRIC APPLICATIONS

Orthogonal Trajectories

A plane curve C is an *orthogonal trajectory* to a family of plane curves P if at each point where C cuts a P-curve, the curves are orthogonal; i.e., the tangent to C is perpendicular to the tangent to the P-curve. Two families of curves, P and Q are said to be *orthogonal families* if each P-curve is an orthogonal trajectory to the family Q and vice versa.

Orthogonal trajectories occur frequently in physical applications. In each of the following contexts, the notion of an orthogonal family has a physical interpretation:

(a) electricity and magnetism if P denotes a family of curves of constant electric or magnetic potential (the so called *equipotential curves*) then the orthogonal family Q represents the lines of electric or magnetic force.

(b) fluid flow if P denotes a family of curves of constant fluid potential then the orthogonal family Q is the family of *streamlines*. These are the trajectories followed by fluid particles in the flow.

(c) heat conduction if P denotes the family of curves of constant temperature (the so called *isotherms*) then the orthogonal family Q indicate the curves along which heat flows.

(d) atmospheric science if P denotes a family of *isobars*, curves of constant pressure, then the orthogonal family Q indicates the wind flow pattern.

DESCRIPTIONS OF CURVES

A curve or family of curves in the plane may be described in various ways:

a) Explicit description Here the curve is defined to be the set of points (x,y) where one of the variables x or y is given as an explicit function of the other. For example, as the parameter k varies, the equation

$$y = kx$$

describes a family of straight lines all passing through the origin.

b) Implicit description Here the curve is defined to be the set of points where a function, F(x,y), of two variables is equal to a fixed constant. Changing the value of the constant generally changes the curve. For example, as the parameter R varies, the set of points (x,y) that satisfy

$$x^2 + y^2 = R^2$$

is a family of concentric circles, centered at the origin.

Example 3.1 A Family Of Orthogonal Trajectories

We consider now an example of a family of curves that is described explicitly and show how to find the family of orthogonal trajectories. As the parameter k ranges over all positive values, the equation,

$$y(x) = k/x,$$

describes a family of hyperbolas having the coordinate axes as asymptotes. Along any curve in this family we have

$$y'(x) = -k/x^2.$$

Eliminating the parameter k between these two equations leads to

$$y' = -(yx)/x^2 = -y/x.$$

Here y' is the slope of the tangent line at any point (x,y) along any one of the hyperbolas. Then at a point (x,Y) on an orthogonal trajectory, the slope of the tangent must be equal to the negative reciprocal of y'; i.e., if the orthogonal trajectory is given by $Y = Y(x)$, then

$$Y' = x/Y.$$

Solving this differential equation gives

$$\tfrac{1}{2} Y(x)^2 = \tfrac{1}{2} x^2 + c; \text{ so}$$

$$Y^2 - x^2 = C.$$

Thus, the orthogonal trajectories to the family of hyperbolas whose asymptotes are the coordinate axes is a second family of hyperbolas asymptotic to the lines $y = \pm x$. If the first family of hyperbolas consists of the isobars on a weather map, then the second family of hyperbolas consists of the resulting wind trajectories. See Figure 3.2.

SOLVED PROBLEMS

Growth and Decay

PROBLEM 3.1

In a laboratory experiment, it is determined that it takes 80.3 days for the amount of a given radioactive material to decay to .99 of the amount originally present. Find the rate constant k and the half life of this element.

SOLUTION 3.1

Note that 80.3 days is .22 years. Then, using (3.2) we have

$$.99A(0) = A(0)\exp(-.22\,k).$$

Dividing by A(0) and taking the natural log of both sides leads to

$$\ln(.99) = -.22\,k.$$

Then

$$k = .04568$$

and the half-life T is then equal to

$$T = \ln 2/.04568 = 15.17 \text{ years.}$$

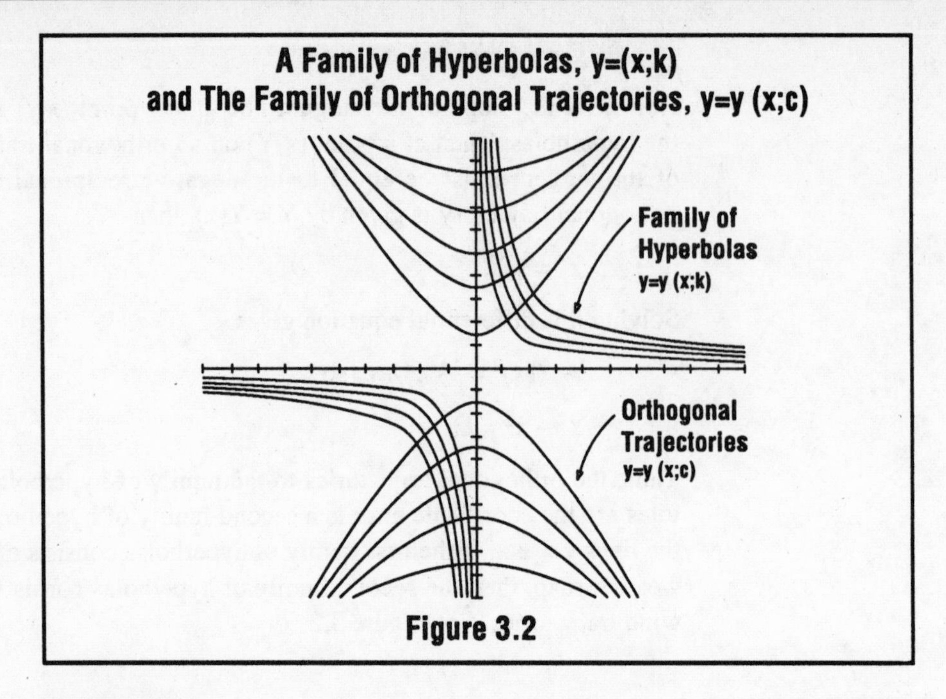

Figure 3.2

PROBLEM 3.2

Carbon 14, a radioactive isotope of carbon, is commonly used for the purpose of dating archeological finds. It is possible to do this since researchers are able to measure the amount of Carbon 14 present in samples found at an archeological site and determine the percentage this represents of the original amount. The half-life of Carbon 14 is 5745 years. What is the age of a relic that contains 20 percent of the original amount of Carbon 14?

SOLUTION 3.2

Since the half-life of Carbon 14 is 5745 years, the rate constant is

$$k = \frac{\ln 2}{5745} = .0001206.$$

Then the time t at which only 20 percent of the original amount of Carbon 14 remains satisfies

$$.20 = \exp(-.0001206t); \text{ so}$$

$$t = \frac{\ln .20}{.0001206} = 13{,}345 \text{ years.}$$

PROBLEM 3.3(A)

Over a short period of time, the growth of a certain bacterial culture is assumed to be governed by equation (3.4) with a constant growth rate k. If the size of the population is observed to double in T hours, then how long (expressed in terms of T) does it take for the population to triple in size? How long to reach four times its original size?

PROBLEM 3.3(B)

Over a longer period of time it is more realistic to assume a growth rate dependent on P(t) as expressed by

$$k = K\,(P^* - P(t))$$

for K,P* given, positive constants. Show that as t increases, P(t) tends to the value P*, independent of whether the initial value, P_0 is greater than or less than P*.

PROBLEM 3.3(C)

If the initial population size P_0 equals .1 P*, then find the time T>0 such that P(T) = .9 P*; the time T will be expressed in terms of K and P*.

SOLUTION 3.3(A)

The solution to (3.4) in the case of constant growth rate k is given by

$$P(t) = P_0 e^{kt}. \tag{1}$$

Then the doubling time, T, satisfies

$$P(T) = 2P_0 = P_0 e^{kT},$$

hence,

$$k = (\ln 2)/T\,. \tag{2}$$

Then substituting this expression for k into (1) leads to

$$P(t) = P_0 (e^{\mathrm{Ln}\,2})^{t/T} = P_0 (2)^{t/T}\,.$$

Clearly, when t=2T, we have P(t) = $4P_0$. Moreover, the time τ for which $P(\tau) = 3P_0$ must then lie between T and 2T. In particular, τ satisfies,

$$3P_0 = P_0 (2)^{\tau/T},$$

or

$$\tau = T(\ln 3/\ln 2) = 1.585T.$$

SOLUTION 3.3(B)

In the case of variable growth rate as expressed in (3.5), the equation is nonlinear but separable, and the solution is a slight modification of the solution to exercise 2.9. The solution may be written as

$$P(t) = \frac{P^* P_0}{P_0 + (P^* - P_0)e^{-P^*Kt}} \tag{3}$$

Clearly, this reduces to P_0 at t=0 and since $P_0 + (P^* - P_0)e^{-P^*Kt}$ tends toward P_0 as t tends to infinity, P(t) tends to the limit P* with increasing t as shown in Figure 3.3.

Note that if P_0 is less than P*, then $P_0 + (P^* - P_0)e^{-P^*Kt}$ decreases toward P_0, while if P_0 is greater than P*, the expression increases toward P_0. Then P(t) does just the reverse, increasing toward P* if P_0 is less than P* and decreasing to P* when P_0 is greater than P*. The parameter P* may be interpreted as the maximum sustainable population size.

SOLUTION 3.3(C)

If $P_0 = .1P^*$, then (3) reduces to

$$P(t) = \frac{.1P^*}{.1 + .9e^{-P^*Kt}} \tag{4}$$

and if $P(T) = .9P^*$, then T must satisfy

$$.9 = \frac{1}{1 + 9e^{-P^*KT}} \tag{5}$$

Solving (5) for T, we obtain $T = \ln 81/P^*K$. Here T is the time required for the population to grow from ten percent to ninety percent of the maximum sustainable population size.

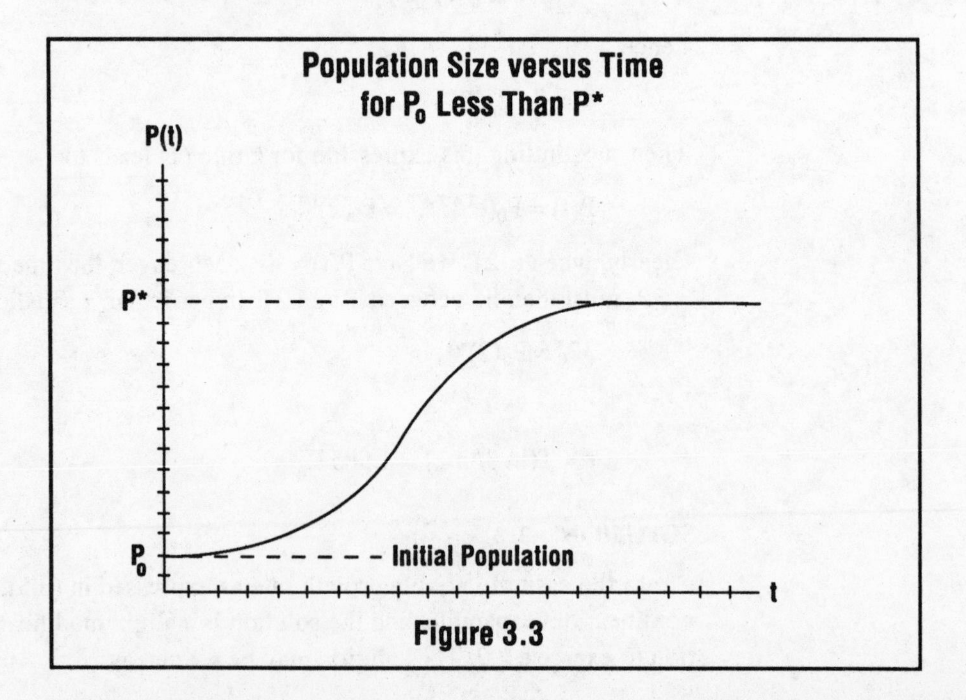

Figure 3.3

Heat Transfer

PROBLEM 3.4

A steel bearing that has been heated to a temperature of 300° C is dropped into a liquid bath whose temperature is 20° C. After 40 seconds, the temperature of the bearing has dropped to 292° C. What will the temperature of the bearing equal after ten minutes, assuming that the bath is large enough that the bearing does not heat the bath significantly; i.e., the temperature of the bath remains constant.

SOLUTION 3.4

If we denote the temperature of the bearing by $T(t)$ and let T^* denote the temperature of the bath, then $T(t)$ satisfies equation (3.6). If we let $S(t) = T(t) - T^*$, then $S'(t) = T'(t)$ and

$$S'(t) = -kS(t).$$

It follows that

$$S(t) = Ce^{-kt}; \text{ so}$$

$$T(t) = T^* + Ce^{-kt} = 20 + Ce^{-kt}.$$

Now

$$T(0) = 20 + C = 300° \text{ C},$$

hence

$$T(t) = 20 + 280e^{-kt}.$$

We are given that

$$T(40) = 20 + 280e^{-40k} = 292,$$

and we can solve this equation for k to obtain

$$k = \ln(280/272)/40 = 7.2468 \text{ X } 10^{-4}.$$

Now, after ten minutes (600 seconds), we have

$$T(600) = 20 + 280e^{-.4347} = 201.268° \text{ C}.$$

PROBLEM 3.5

An air bubble is suspended in a fluid whose temperature, T^*, is varying periodically as follows, $T^* = T_0 \text{ Sin } \Omega t$. Then the temperature, $T(t)$, of the air in the bubble satisfies (3.6). If $T(0) = T_1$, find $T(t)$ and show that the "steady state" temperature of the air in the bubble is periodic but is out of phase with the fluid temperature.

SOLUTION 3.5

The temperature T(t) solves the following problem,

$$T'(t) = -k(T(t) - T_0 \text{Sin } \Omega t)$$

$$T(0) = T_1.$$

According to Theorem 2.1, the solution to this initial value problem is given by

$$T(t) = T_1 e^{-kt} + kT_0 \int_0^t e^{-k(t-\tau)} \text{Sin } \Omega\tau \, d\tau$$

$$= T_1 e^{-kt} + \frac{kT_0}{k^2 + \Omega^2} (k\text{Sin } \Omega t - \Omega \text{Cos } \Omega t + \Omega e^{-kt})$$

If we let

$$\theta = \text{Arcsin}(k/(k^2 + \Omega^2))$$

then the solution can be written

$$T(t) = e^{-kt}(T_1 + \Omega k T_0/(k^2 + \Omega^2)) + kT_0(\text{Cos } \theta \text{ Sin } \Omega t - \text{Sin } \theta \text{ Cos } \Omega t)$$

$$T(t) = e^{-kt} [\, T_1 + \Omega k T_0/(k^2 + \Omega^2)\,] + kT_0 \text{ Sin}(\Omega t - \theta).$$

Steady State Solution Since e^{-kt} tends to zero with increasing t, the solution T(t) tends toward kT_0 Sin(Ωt–θ) as t tends to infinity. This is the so–called *steady–state* solution, i.e., the part of the solution that remains after the *transient* solution has died out. The expression $e^{-kt}(T_1 + \Omega kT_0/(k^2 + \Omega^2))$ is the transient solution; the part of the solution that is due to the initial condition. As time goes on, this component of the solution diminishes in importance; i.e. its effect is transient.

Note that the steady–state solution is periodic with frequency equal to the frequency of the oscillating fluid temperature. Note also that since θ is not zero, the bubble temperature is out of phase with the fluid temperature.

Chemical Mixing

PROBLEM 3.6

A mixing tank whose volume is equal to V is initially filled with a mixture that is 80 percent chemical X. A 20 percent mixture of chemical X is pumped into the tank at the rate of Q gallons per minute and becomes instantly mixed with the fluid already there. At the same time Q gallons per minute of this new mixture are pumped out of the tank. How long will it take for the concentration level of chemical X in the tank to fall to 30 percent? How long will it take for the concentration level to fall to 20 percent?

SOLUTION 3.6

If X(t) denotes the amount of chemical X in the tank at time t, then X(t) satisfies equation (3.7) for f(t) equal to the constant value .20. In addition, X(0)= .80V and hence, according to Theorem 2.1,

$$X(t) = .80V\, e^{-kt} + .20\, Q\, e^{-kt} \int_0^t e^{-k\tau}\, d\tau$$

$$= .20\, V + .60\, V\, e^{-kt},$$

where k = Q/V. In order to find T>0 such that X(T) = .30 V, we solve the following equation for T,

$$.20V + .60Ve^{-kT} = .30V.$$

This leads to

$$T = \frac{\ln 6}{k}.$$

Note that while the concentration level *tends* toward 20 percent as t increases, there is no finite time at which the level actually *equals* 20 percent. More generally, for an initial concentration level of C_1 and an input concentration level of C_2, we find

$$X(t) = C_2V + (C_1 - C_2)Ve^{-kt}.$$

The concentration level assumes every value *between* C_1 and C_2 at a finite (and computable) value of t, but does not *equal* C_2 at any finite time.

PROBLEM 3.7

Fluid is pumped from a mixing tank at the rate of Q gallons per minute. Simultaneously, fluid is being pumped into the tank at the same rate so that the total volume of fluid in the tank remains equal to the constant value V. The incoming fluid contains a chemical X, for which the concentration level, f(t), varies with time according to

$$f(t) = \begin{cases} 1 & \text{if } 0 < t < 1 \text{ min} \\ 0 & \text{if } 1 < t \end{cases} \tag{1}$$

Find X(t), the amount of chemical X in the tank at time t ≥ 0, if the tank is initially devoid of chemical X.

When does the amount of chemical X in the tank reach a maximum and what is this maximum value. At what time T > 1 does it happen that X(T) equals half the maximum value.

SOLUTION 3.7

X(t) satisfies equation (3.7) and, since there is no chemical X in the tank to start with, we have the initial condition X(0)=0. This is a linear, inhomogeneous equation and, according to Theorem 2.1, the solution is given by

$$X(t) = Q \int_0^t e^{-k(t-\tau)} f(\tau)\, d\tau \tag{2}$$

where we have let $k = Q/V$. Observe that the rate constant in this reaction is determined by how fast the pumps can change the fluid in the mixing tank.

Using (1) in (2) leads to

$$X(t) = \begin{cases} Q \int_0^t e^{-k(t-\tau)}\, d\tau & \text{if } 0 < t < 1, \\ Q \int_0^1 e^{-k(t-\tau)}\, d\tau & \text{if } t > 1. \end{cases}$$

That is,

$$X(t) = \begin{cases} V(1 - e^{-kt}) & \text{if } 0 < t < 1, \\ V(e^{-k(t-1)} - e^{-kt}) & \text{if } t > 1\,. \end{cases}$$

$X(t)$ is plotted versus t in Figure 3.4. It can be clearly seen that $X(t)$ is a continuous function whose first derivative is discontinuous at $t=1$. $X(t)$ increases from the value zero at $t=0$ until $t=1$ where it achieves its maximum value of $V(1-e^{-k})$. To find $T>1$ such that $X(T)$ equals one half this maximum value, we solve

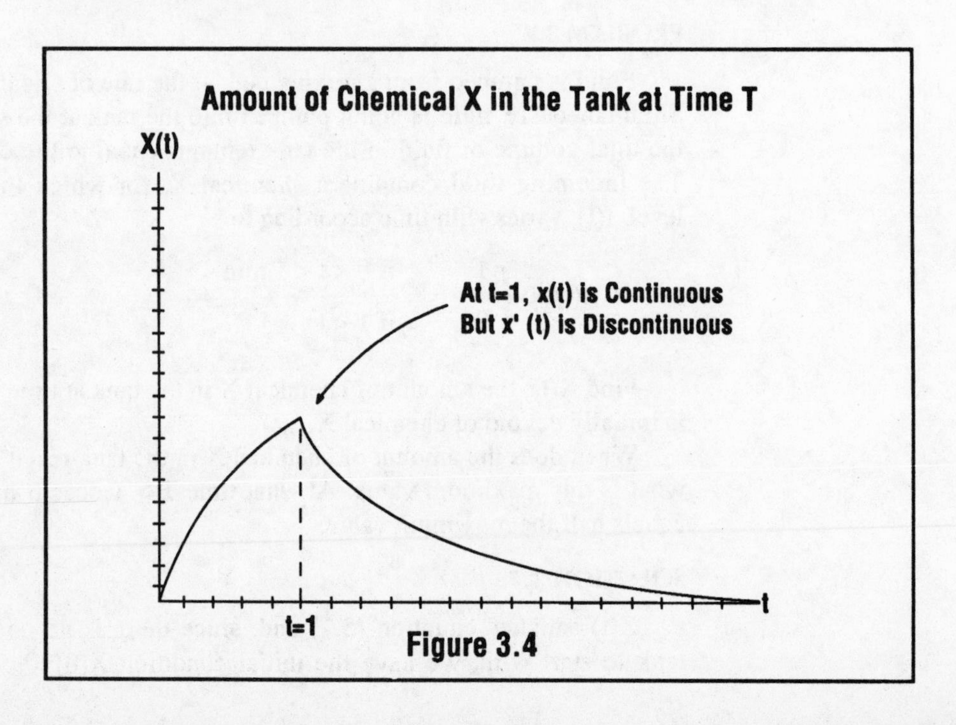

Figure 3.4

$$V(e^{-k(T-1)} - e^{-kT}) = \tfrac{1}{2} V(1-e^{-k}); \text{ so}$$

$$e^{-kT}(e^{k} - 1) = \tfrac{1}{2}(1-e^{-k}).$$

Solving this last equation for T, we obtain the result,

$$T = 1 + (\ln 2)/k.$$

More generally, it is easy to see that for any positive integer N, X(T) is equal to $V(1-e^{-k})/N$ (i.e. 1/N times the maximum value) at the time,

$$T = 1 + (\ln N)/k\,.$$

Thus while the amount of chemical X in the tank theoretically is never zero after t=0, we can compute the time at which the amount is equal to $V(1-e^{-k})/N$ for every N.

Mechanics

PROBLEM 3.8

A small mass is dropped into a viscous fluid, where it falls under its own weight opposed only by the viscous drag of the fluid. If the viscous drag is proportional to the square of the velocity of the mass, then show that the velocity of the mass does not continue to increase as the mass falls, but reaches a maximum or "terminal" velocity that depends on the weight of the mass and on the viscous drag.

SOLUTION 3.8

If the mass of the object is equal to m and the velocity is denoted by v(t), then Newton's law requires

$$d/dt\,(mv) = F,$$

where F denotes the net force acting on the object. In this case, the forces acting on the falling mass are the force of gravity (equal to the weight of the object) and the drag force; i.e.,

$$F = mg - D\,v^2,$$

where D is a positive constant expressing the proportionality between the square of the velocity and the drag force. The larger the number D, the larger is the drag force opposing the downward motion of the falling object. Note that F depends explicitly on v but not on t or on the position x.

Here we are supposing that the density of the fluid is negligible with respect to the density of the falling object and hence there is no "bouyant" force opposing the fall of the object. In the event that the fluid density is not negligible then we may include the bouyant force of the fluid in this analysis by replacing the expression mg (the weight of the object) by

$$(d_1 - d_2)Wg = (m - m')g\,.$$

Here d_1 and d_2 denote the density of the object and the fluid, respectively, and the volume of the object is equal to W. Thus, mg is the weight of the object and m′g is the weight of the displaced volume of fluid.

Here we shall continue to assume the bouyant force is negligible. If we assume that the object is released from rest in the viscous fluid, we have the following initial value problem for v(t):

$$d/dt\,(mv) = mg - Dv^2, \; v(0) = 0. \tag{1}$$

This is a nonlinear first order equation in which the variables separate as follows,

$$\int \frac{dv}{Q^2 - v^2} = (D/m) \int dt$$

Here we have let $Q^2 = mg/D$. Then

$$\frac{1}{2Q} \int \left(\frac{1}{v+Q} - \frac{1}{v-Q} \right) dv = (D/m) \int dt$$

and

$$\ln\left(\frac{v+Q}{v-Q}\right) = 2kt + c,$$

where $k = QD/m = (Dg/m)^{1/2}$. Solving for v in terms of t leads to

$$v(t) = Q\,\frac{C + e^{-2kt}}{C - e^{-2kt}}\,.$$

In order that v(0)=0, we must have C = –1, and hence

$$v(t) = Q\,\frac{1 - e^{-2kt}}{1 + e^{-2kt}}\,. \tag{2}$$

Now as t tends to infinity, v(t) tends to the value $Q = (mg/D)^{1/2}$. This is the terminal velocity achieved by the falling object. Note that this terminal velocity is directly proportional to the square root of the object's weight and is inversely proportional to the square root of the drag parameter.

Since v(t) = dx/dt, we could now proceed solve (2) to find the position x as a function of t. This would be necessary, for example, if the depth of the fluid were finite and we wanted to know the velocity of the falling object at the time it strikes the bottom.

PROBLEM 3.9

A projectile is shot upward from the surface of the Earth with an initial velocity equal to V_0. The rise of the projectile is opposed only by the gravitational attraction of the Earth. Find the smallest value of the initial velocity, V_0, in order that the projectile will eventually escape the Earth's gravitational field.

SOLUTION 3.9

Let x denote the position of the projectile, with x equal to zero at the surface of the Earth and increasing in the upward direction as seen in Figure 3.5. Then an upward velocity is positive and the attraction of the Earth acting on the projectile is negative. The magnitude of this attractive force is proportional to the product of the masses of the Earth and the projectile and inversely proportional to the square of the distance of the projectile from the center of the Earth; i.e.

$$F = -G \frac{M\,m}{(R+x)^2} .$$

Here M, R denote the mass and radius of the Earth, respectively, and G denotes the constant of proportionality. Note that since F must equal mg at the surface of the Earth (where x=0), we have

$$G = g\,R^2/M$$

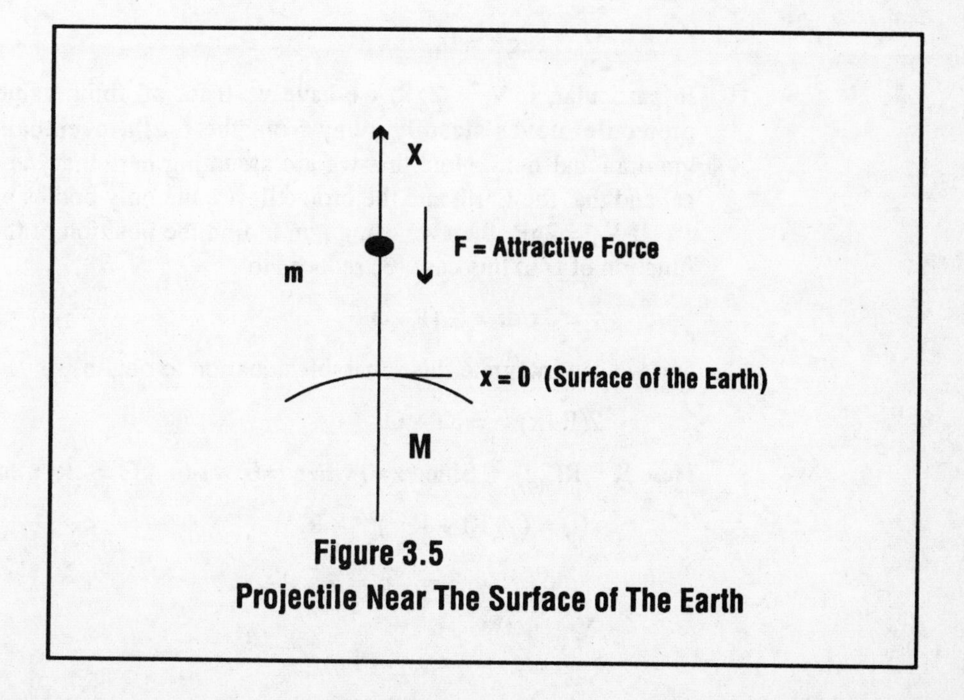

Figure 3.5
Projectile Near The Surface of The Earth

Hence (3.8) becomes in this application,

$$d/dt\,(mv) = -\frac{mgR^2}{(R+x)^2}\,. \tag{1}$$

Since the right side of (1) depends explicitly on x and not on t, we use the form (3.9) of the equation of motion for the projectile. Then $v=v(x)$ satisfies

$$v\,dv/dx = -\frac{gR^2}{(R+x)^2}\,, \quad v(0) = V_0 \tag{2}$$

The nonlinear equation is separable, and we integrate to find

$$\tfrac{1}{2}v^2 = \frac{gR^2}{R+x} + C\,.$$

The initial conditionv$=V_0$ when $x=t=0$, implies

$$\tfrac{1}{2}V_0^2 = gR + C$$

hence,

$$v^2 = V_0^2 - 2gR + \frac{2gR^2}{R+x} \tag{3}$$

It is evident from (3) that in order to have $v^2>0$ for all $x>0$, it is necessary that

$$V_0^2 - 2gR > 0.$$

In particular, if $V_0^2= 2gR$, we have $v^2>0$ for all finite values of x, and the projectile moves steadily away from the Earth, eventually escaping the gravitational field. Note that we are assuming here that the Earth is spherical and that the Earth and the projectile are the only bodies present.

If $V_0^2= 2gR$, then we can go on to find the position of the projectile as a function of t. In this case (3) reduces to

$$v = dx/dt = K/(R+x)^{1/2},$$

and we can integrate this separable equation to obtain

$$2(R+x)^{3/2} = Kt + C.$$

Here $K = R(2g)^{1/2}$. Since $x=0$ when $t=0$, we find $C = 2R^{3/2}$, hence,

$$x(t) = (\tfrac{1}{2}Kt + R^{3/2})^{2/3} - R.$$

PROBLEM 3.10

A ball of mass m is attached to one end of a rigid rod of length L that pivots freely about the other end (Figure 3.6). We suppose the mass of the rod is negligible and that the ball pivots in a fixed plane. Denote the angle the rod makes with the vertical by θ, and then write the equation of motion in terms of θ for the ball executing a pendulum motion under its own weight.

(a) Assume that the pivot is frictionless and show that the sum of the kinetic and potential energies of the ball is constant in time.

(b) Consider the case in which the pivot exerts a resistive force whose tangential component is proportional to $\theta'(t)$, and show that then the total energy is a steadily decreasing function of time.

SOLUTION 3.10(A)

Since the ball moves in a plane, the velocity of the ball and the forces acting on the ball all have two components. We can choose to resolve these quantities into their radial and tangential components; since the rod is rigid, the radial component of velocity is equal to zero. The tangential component of the velocity is equal to $L\theta'(t)$, and the tangential component of the ball's weight is equal to $mg\text{Sin}\,\theta$.

Equation (3.9) applies to the tangential components of the momentum and the force. That is, if the pivot provides no resistance, then

$$d/dt\,(mL\theta') = -\,mg\text{Sin}\,\theta. \qquad (1)$$

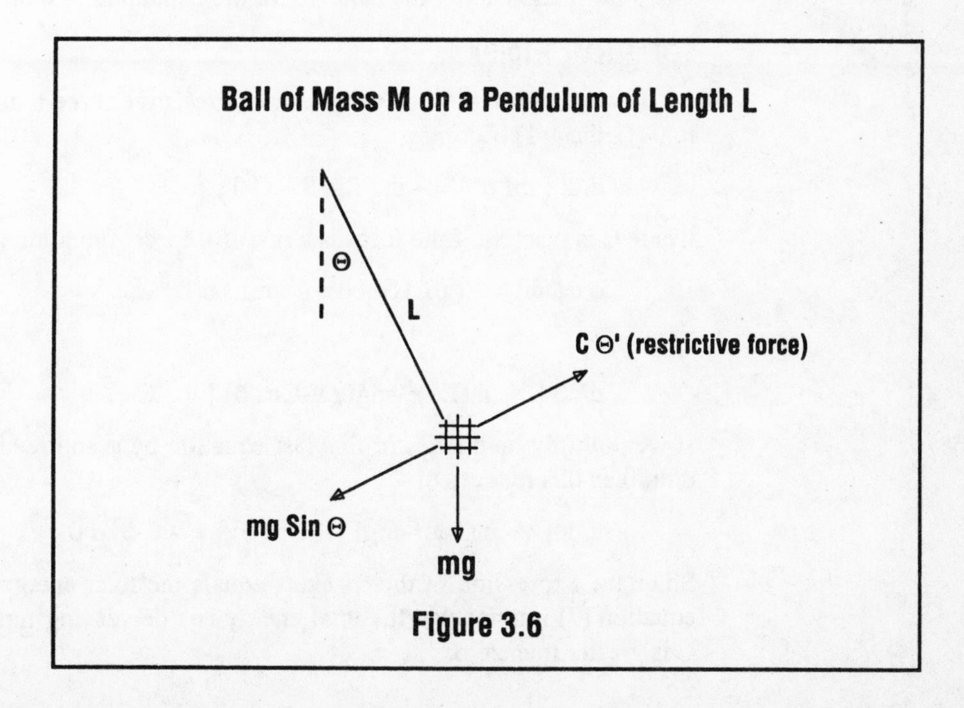

Figure 3.6

Since the right side of (1) depends explicitly on θ but not on t, we let $\phi = d\theta/dt$ and write,

$$d/dt\ \phi\ (\theta) = d\phi/d\theta\ d\theta/dt = \phi\ d\phi/d\theta.$$

Then (1) becomes

$$\phi\ d\phi/d\theta = -\ (g/L)\ \text{Sin}\ \theta\ . \tag{2}$$

This is a separable nonlinear equation for $\phi=\phi(\theta)$. We can integrate (2) to obtain

$$\tfrac{1}{2}\ \phi(\theta)^2 - \tfrac{1}{2}\ \phi(0)^2 = (g/L)(1 - \text{Cos}\ \theta). \tag{3}$$

We can rewrite this as

$$\tfrac{1}{2}\ m\ L^2\ \phi^2 - mgL(1-\text{Cos}\ \theta) = \tfrac{1}{2}\ mL^2\ \phi_0^2 \tag{4}$$

Observe that since

$$\tfrac{1}{2}\ mL^2\ \phi^2 = \text{kinetic energy of the ball, and}$$

$$-\ mgL(1-\text{Cos}\ \theta) = \text{potential energy of the ball,}$$

equation (4) expresses the fact that the sum of the kinetic and potential energies is equal to a constant.

The expression (4) is referred to as a *FIRST INTEGRAL* for the differential equation (2). It does not provide an explicit formula for θ (or ϕ) in terms of t, so it is not a solution of the equation. Nevertheless, it does provide information about the behavior of the pendulum system.

SOLUTION 3.10(B)

If we suppose that the pivot exerts a resistive force that is proportional to $\theta'(t)$, then (1) becomes

$$d/dt\ (\ mL\theta'\ ) = -\ mg\ \text{Sin}\ \theta - C\ \theta', \tag{5}$$

where C is positive if the force is a *resistive* force. Equation (2) changes to

$$\phi\ d\phi/d\theta = -\ (g/L)\ \text{Sin}\ \theta - (C/mL)\ \theta'. \tag{6}$$

That is,

$$d/d\theta\ [\ \tfrac{1}{2}\ m(L\phi)^2 - mgL(1-\text{Cos}\ \theta)\ ] = -\ C\ \phi.$$

If we multiply both sides of this last equation by ϕ and recall that $\phi\ d/d\theta = d/dt$, then this reduces to

$$d/dt\ [\ \tfrac{1}{2}\ m(L\phi)^2 - mgL(1-\text{Cos}\ \theta)\ ] = -\ C\ \phi^2 \le 0 \tag{7}$$

Since the expression in the brackets equals the total energy of the system, equation (7) implies that the total energy is a decreasing function of time if C is greater than zero.

Geometry

PROBLEM 3.11

Find the orthogonal trajectories to the family of circles described implicitly by,

$$(x-c)^2 + y^2 = c^2 \qquad c = \text{real parameter}. \tag{1}$$

SOLUTION 3.11

Equation (1) describes a family of circles having centers on the x–axis and tangent to the y–axis at the origin. On any circle in this family we have

$$2(x-c) + 2y\, y' = 0 \tag{2}$$

and if we use (1) to eliminate the parameter c from (2), then

$$y'(x) = (y^2-x^2)/2xy. \tag{3}$$

If $z=z(x)$ denotes the explicit representation for the orthogonal family, then it follows from (3) that

$$z'(x) = -\,2xz/(z^2-x^2). \tag{4}$$

Equation (4) is a nonlinear differential equation in which the function

$$F(x,z) = -\,2xz/(z^2-x^2)$$

satisfies $F(x,z) = F(kx,kz)$; i.e., F is homogeneous of degree zero. Then if we let

$$z(x) = x\, v(x),$$

we find

$$x\, v'(x) + v(x) = 2v/(1-v^2).$$

This equation is separable with the solution

$$v/(1+v^2) = Cx\,;$$

that is,

$$z = C(x^2+z^2).$$

We may rewrite this as

$$x^2 + (z-b)^2 = b^2,$$

for $b = 1/2C$. This is a family of circle having centers on the vertical axis and tangent to the x–axis at the origin. The two families of circles are shown in Figure 3.7.

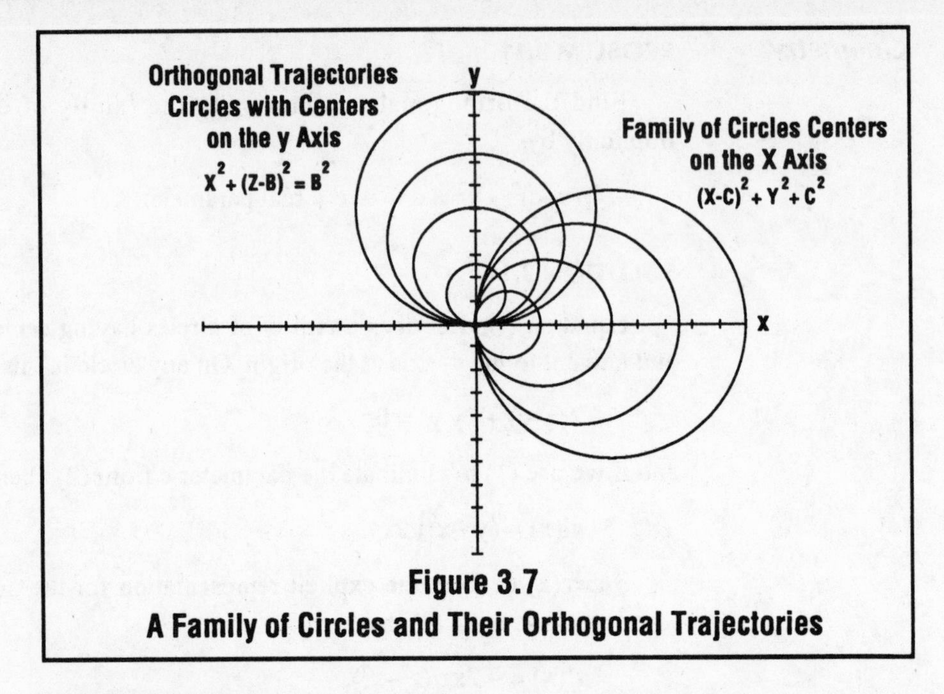

Figure 3.7
A Family of Circles and Their Orthogonal Trajectories

PROBLEM 3.12

A point electric charge is located at the origin in the plane. Surrounding this point charge are concentric circles denoting the equipotential curves. Find the "lines of electric force" generated by this point charge.

SOLUTION 3.12

The lines of force are the orthogonal trajectories of the equipotential curves represented by the family of circles. We can describe these circles by the equation

$$x^2 + y^2 = R^2. \tag{1}$$

Then along any of these concentric circles,

$$2x + 2y\, y' = 0. \tag{2}$$

Note that since the curves in this family are concentric circles, we can solve (2) for $y'(x)$ at any point (x,y) on any circle in the family without using (1) to eliminate the parameter R. Thus,

$$y'(x) = -x/y, \tag{3}$$

and along the orthogonal trajectory we have

$$z'(x) = z/x.$$

The solution of this equation is given by

$$z(x) = Cx.$$

This family of straight lines through the origin indicates the lines of force corresponding to the equipotential curves of (1). The lines and curves are shown in Figure 3.8.

PROBLEM 3.13

It has been hypothesized that when spherical pellets of chemical X are dropped into a solution of chemical Y, the pellets dissolve at a rate that is proportional to the P–th power of the exposed surface area. Assuming that the pellets maintain their spherical shape as they dissolve, devise an experiment to determine the value of the parameter p, given that p is greater than or equal to 2.

SOLUTION 3.13

The hypothesis states the the pellets dissolve at a rate proportional to the Pth power of the exposed surface area. If we denote the volume and surface area of the pellet by V and A, respectively, then the hypothesis is equivalent to

$$dV/dt = -k A^P. \qquad (1)$$

Here k denotes the positive constant of proportionality and the negative sign in (1) reflects the fact that the volume of the pellet is decreasing with time (the pellet is dissolving).

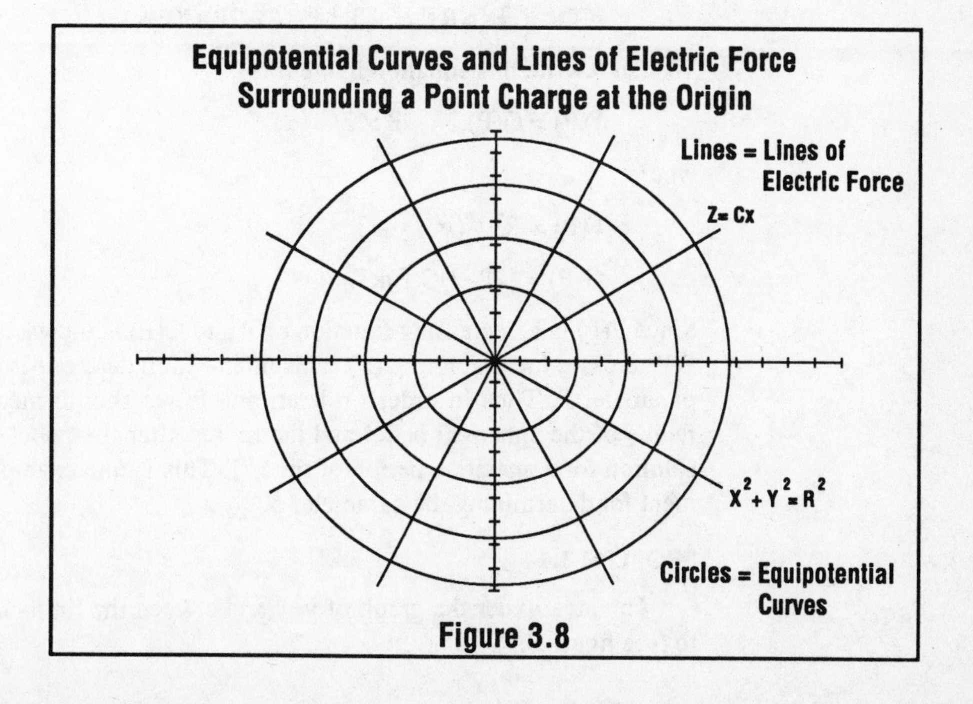

Figure 3.8

Assuming that the pellets maintain their spherical shape as they dissolve, we make use of the formulas

$$V = 4\pi r^3/3, \qquad A = 4\pi r^2,$$

to write,

$$d/dt(4\pi r^3/3) = -\,k(4\pi r^2)^P\,;$$

i.e.

$$dr/dt = -\,kC(P)r^{2P-2} \tag{2}$$

where

$$C(P) = (4\pi)^P/12\pi.$$

Equation (2) separates and yields the solution,

$$r(t)^{3-2P} = D - (3 - 2p)C(P)kt, \tag{3}$$

where D denotes a constant of integration.

In order to determine the parameter P, suppose that after spending a time T in the solution, a spherical pellet whose radius was initially equal to $R>0$ has shrunk to a pellet of radius $\propto R$ for $\propto<1$. Then it follows from (3) that

$$D = R^{3-2P},$$

and

$$r(T)^{3-2P} = (\propto R)^{3-2P} = R^{3-2P} - (3-2P)C(P)kT.$$

We can rewrite this equation in the form,

$$F(P) = G(P), \qquad P > 2$$

where

$$F(P) = R^{3-2P}(\propto^{3-2P} - 1)$$

$$G(P) = (2P-3)C(P)kT.$$

Since F(P) is a decreasing function of P and G(P) is increasing, the graph of F(P) crosses the graph of G(P). This intersection determines the value of the parameter P. Thus in order to determine P, we should measure the initial radius of the spherical pellet and the radius after the pellet has been in the solution for a specified period of time, T. This is one example of an experiment for determining the parameter p.

PROBLEM 3.14

The area under the graph of $y=y(x)$ between the limits 0 and x is equal to (see figure 3.9)

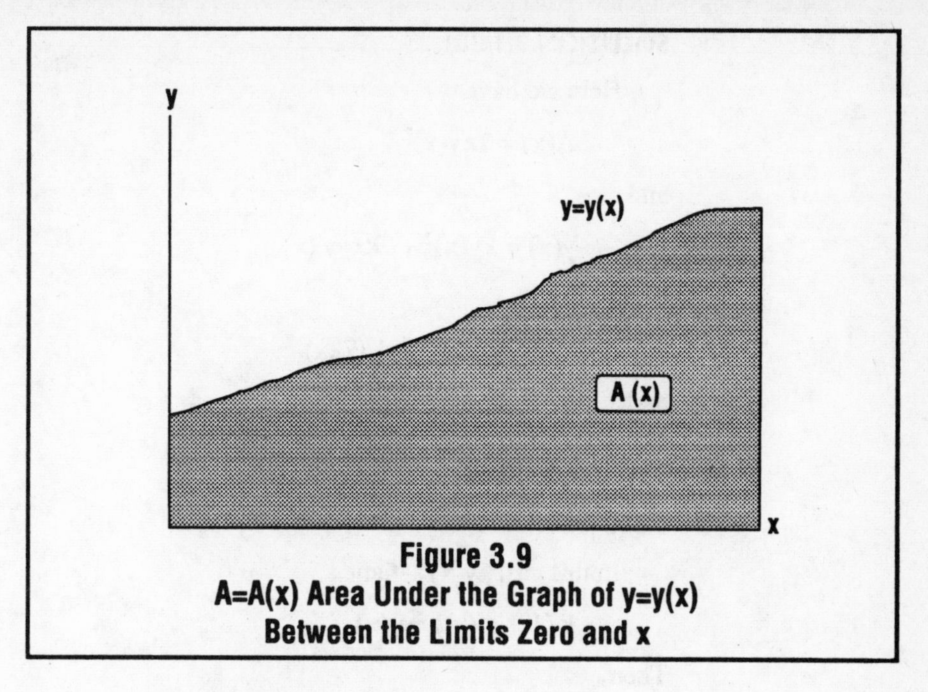

Figure 3.9
A=A(x) Area Under the Graph of y=y(x) Between the Limits Zero and x

$$A(x) = \int_0^x y(t)\,dt\,.$$

Find the function y=y(x) such that:

(a) A(x) is proportional to xy(x).

(b) A(x) is proportional to $xy(x)^2$

(c) A(x) is proportional to x + y(x).

(d) A(x) is proportional to y(x)/x .

SOLUTION 3.14(A)

Note that $A'(x) = y(x)$. If A(x) is proportional to xy(x), then for some constant k,

$$A(x) = k\,x\,y(x);$$

differentiating both sides,

$$A'(x) = y(x) = ky(x) + kxy'(x).$$

Then

$$kxy'(x) = (1-k)y(x)$$

and

$$y(x) = Cx^b \qquad b = (1-k)/k.$$

SOLUTION 3.14(B)

Here we have

$$A(x) = kxy(x)^2$$

and

$$y(x) = ky(x)^2 + 2kxyy'(x).$$

Then

$$y'(x) + y/(2x) = 1/(2kx),$$

and

$$y(x) = 1/k + Cx^{-1/2}.$$

SOLUTION 3.14(C)

In this case, y(x) satisfies

$$k\,(1 + y'(x)) = y(x)\,.$$

Then

$$y(x) = Ce^{x/k} + k.$$

SOLUTION 3.14(D)

In order for A(x) to be proportional to y/x, y(x) must satisfy

$$k(y'(x)/x - y/x^2) = y;\ \text{so}$$

$$y' - (x/k + 1/x)y = 0.$$

Then

$$y(x) = Cx\,\exp(x^2/2k).$$

In this chapter we have surveyed several applications involving first order differential equations. Some of these applications derive from situations in which the rate of change of the dependent variable is proportional to the present value of that variable. These applications include

- *RADIOACTIVE DECAY*
- *POPULATION MODELS*
- *NEWTON'S LAW OF COOLING*
- *CHEMICAL MIXING*

In the case of chemical mixing, the rate at which the amount of chemical present increases is proportional to the difference between inflow and outflow rates. However, inflow and outflow rates are proportional to the amount of chemical currently present.

Knowledge of the solution to the radioactive decay equation makes possible such things as carbon dating of archeological data. In connection with the other applications in this category it is often desirable to use the solution to the equation for the purpose of predicting the asymptotic behavior of the dependent variable.

Newton's second law which states that the momentum of a body changes at a rate proportional to the net force acting on the body leads also to differential equations of order one. Often the equations obtained in this connection are nonlinear but even when the equations are not explicitly solvable useful information may be obtained from a first integral.

Finally, there are a variety of geometrical applications that involve first order equations. For example there are many physical applications in which it is useful to be able to compute the orthogonal trajectories to a given family of curves. Other geometric examples include problems based on the fact that the volume of a dissolving solid pellet decreases at a rate proportional to the current value of its surface area.

4

Higher Order Equations

Chapters two and three are devoted to the solution and applications of differential equations of order one. Of course equations of order higher than one are of interest as well.

We begin this chapter with a discussion of general solutions of the general linear *differential equation of order N. In the case of constant coefficient linear equations we shall show how these general solutions may be constructed. Following this, we consider the construction of* particular solutions *for inhomogeneous linear equations. We present two techniques for constructing particular solutions. Each of these solution methods is, in principle, feasible for linear equations with variable coefficients but is generally practical only for constant coefficient equations.*

LINEAR EQUATIONS OF ORDER N

The most general differential equation of order N is written

$$F(x,y,y',...,y^{(N)}) = 0 \tag{4.1}$$

However when the equation is stated in this generality we are unable to say very much about the existence of solutions and still less about their construction. Therefore we shall restrict our attention to *linear equations of order N*. The most general *linear* differential equation of order N may be written as

$$L[y(x)] = f(x) \tag{4.2}$$

where we have used the operator notation , $L[y(x)]$, to denote the expression,

$$L[y(x)] = y^{(N)}(x) + a_{N-1}\, y^{(N-1)}(x) + \dots + a_0\, y(x) \tag{4.3}$$

The coefficients $a_0,\dots,a_{N-1}$ may be functions of the independent variable x. In this case we shall suppose always that they are continuous functions of x.

GENERAL SOLUTIONS

To discuss the solution of (4.2) with efficiency and precision, we must define certain concepts.

LINEARLY DEPENDENT AND INDEPENDENT FUNCTIONS

Functions $y_1(x), y_2(x), \dots, y_M(x)$ are said to be *linearly dependent* if there exist constants $C_1, C_2, \dots, C_M$ not all zero, such that

$$C_1 y_1(x) + \dots + C_M y_M(x) = 0 \qquad \text{for all x.}$$

If the functions $y_1(x), y_2(x), \dots, y_M(x)$ are not linearly dependent then they are said to be *linearly independent*.

WRONSKIAN DETERMINANT

For arbitrary functions $y_1(x), y_2(x), \dots, y_M(x)$ it may not be easy to determine whether they are linearly dependent. However, if the functions $y_1(x), y_2(x), \dots, y_M(x)$ are sufficiently smooth then there is a simple test for dependence.

First we define the *Wronskian Determinant* of the functions $y_1(x), y_2(x), \dots, y_M(x)$.

$$W[y_1, y_2, \dots, y_M](x) = \begin{vmatrix} y_1(x) & y_2(x) & \dots & y_M(x) \\ y_1'(x) & y_2'(x) & \dots & y_M'(x) \\ \cdot & \cdot & & \cdot \\ \cdot & \cdot & & \cdot \\ \cdot & \cdot & & \cdot \\ y_1^{(M-1)} & y_2^{(M-1)} & \dots & y_M^{(M-1)} \end{vmatrix}$$

i.e., for a fixed x we evaluate $y_1(x), y_2(x), \ldots, y_M(x)$ as well as the derivatives through order M–1 of each of these functions. Then the *Wronskian Determinant* $W[y_1, y_2, \ldots, y_M](x)$ is the value obtained when we compute the M by M determinant indicated above.

Example 4.1 Wronskian deteminant

4.1(A)

The Wronskian of the functions

$$y_1(x) = \text{Sin } px, \text{ and } y_2(x) = \text{Cos } px$$

is the following 2 by 2 determinant:

$$W[y_1, y_2](x) = \begin{vmatrix} \text{Sin } px & \text{Cos } px \\ p\text{Cos } px & -p\text{Sin } px \end{vmatrix} = -p$$

4.1(B)

The Wronskian of the functions

$$y_1(x) = x,\ y_2(x) = x^2, \text{ and } y_3(x) = x^3$$

is the 3 by 3 determinant

$$W[y_1, y_2, y_3](x) = \begin{vmatrix} x & x^2 & x^3 \\ 1 & 2x & 3x^2 \\ 0 & 2 & 6x \end{vmatrix} = 2x^3$$

4.1(C)

The Wronskian of the functions

$$y_1(x) = \text{Sin } 2x \text{ and } y_2(x) = \text{Sin } x \text{ Cos } x$$

is the 2 by 2 determinant

$$W[y_1, y_2](x) = \begin{vmatrix} \text{Sin } 2x & \text{Sin } x \text{ Cos } x \\ 2 \text{ Cos } 2x & \text{Cos}^2x - \text{Sin}^2x \end{vmatrix}$$

$$= \text{Sin } 2x\,[\text{Cos}^2x - \text{Sin}^2x] - 2 \text{ Cos } 2x \text{ Sin } x \text{ Cos } x$$

Using the identities

$$\text{Cos}^2x - \text{Sin}^2x \equiv \text{Cos } 2x$$

$$2\text{Sin } x \text{ Cos } x \equiv \text{Sin } 2x,$$

we see that this Wronskian is zero for all values of x.

4.1(D)

The Wronskian of the functions

$$y_1(x) = x^2 \text{ and } y_2(x) = x\,|x|$$

is the 2 by 2 determinant

$$W[y_1,y_2](x) = \begin{vmatrix} x^2 & x\,|x| \\ 2x & 2\,|x| \end{vmatrix} = 0\,.$$

We now have Theorem 4.1, which explains the role of the Wronskian in determining linear dependence of smooth functions.

Theorem 4.1 Linear Dependence and the Wronskian

Theorem 4.1 Suppose the N functions $y_1(x),y_2(x),...,y_N(x)$ are continuous, together with their derivatives of every order up to and including the order N. Then these functions are linearly dependent if and only if the Wronskian $W[y_1,y_2,...,y_N](x)$ is zero for all values of x.

All of the functions in parts (a), (b), and (c) of example 4.1 are continuous and have continuous derivatives of all orders. Then Theorem 4.1 may be applied to decide if these functions are linearly dependent. In parts (a) and (b) of the example, the Wronskian is not zero for all x hence the functions in parts (a) and (b) are linearly independent. The Wronskian in part (c) of the example is zero for all x hence these functions must be dependent. Of course, in view of the identities cited, it is evident that $y_1(x)=2y_2(x)$.

The functions in part (d) of the example are linearly independent yet the Wronskian of these functions is zero for all x. This does not violate Theorem 4.1, however, since these functions are continuous and have continuous first derivatives but the second derivative of y_2 is discontinuous. Therefore, the hypotheses of the theorem are not satisfied, and the theorem does not apply in this case.

If y(x) is a solution of the Nth-order homogeneous equation $L[y(x)] = 0$, where L is given by (4.3) and the coefficients in L are continuous functions of x, then y(x) must be continuous together with all derivatives up to and including the derivative of order N. In particular, Theorem 4.1 applies to such functions; i.e., the solutions are dependent if and only if the Wronskian is zero for all x.

The next theorem explains the importance of linearly independent solutions of the homogeneous equation.

Theorem 4.2 Linearly Independent Solutions

Theorem 4.2 Suppose L is given by (4.3) and the coefficients $a_k=a_k(x)$ $k=0,1,...,N-1$ are continuous functions of x.

(a) Then the homogeneous equation of order N, $L[y(x)] = 0$, has N linearly independent solutions $y_1(x),y_2(x),\ ...,\ y_N(x)$.

(b) Every solution of the homogeneous equation of order N may be written in terms of the solutions $y_1(x), y_2(x), \ldots, y_N(x)$; i.e., it can be written in the form,

$$Y_H(x) = C_1 y_1(x) + C_2 y_2(x) + \ldots + C_N y_N(x)$$

for some choice of the constants $C_1, \ldots, C_N$.

(c) Every solution of the inhomogeneous equation (4.2) can be written in the form

$$Y_P(x) = C_1 y_1(x) + C_2 y_2(x) + \ldots + C_N y_N(x) + y(x)$$

for some choice of the constants $C_1, \ldots, C_N$, and *any* function y(x) which satisfies (4.2).

The solutions of the homogeneous and inhomogeneous Nth–order linear equation, whose existence is asserted by parts (b) and (c) of the preceding theorem, are *general solutions* since they contain N arbitrary constants of integration.

INITIAL VALUE PROBLEM

The function y(x) is said to be a solution of the *initial value problem of order N* if y(x) satisfies the equation (4.2) and in addition satisfies the *initial conditioins,*

$$y(x_0) = Y_0,\ y'(x_0) = Y_1,\ \ldots,\ y^{(N-1)}(x_0) = Y_{N-1} \tag{4.4}$$

Here x_0 and $Y_0, \ldots, Y_{N-1}$ denote specified constants. We can prove the following theorem,

Theorem 4.3 Solvability of the Initial Value Problem

Theorem 4.3 Suppose that the coefficients $a_0, \ldots, a_{N-1}$ are continuous functions of x. For each continuous function f(x) and every set of constants x_0 and $Y_0, \ldots, Y_{N-1}$ there exists a unique solution of the initial value problem (4.2), (4.4).

Although Theorem 4.3 asserts the existence of a unique solution to the initial value problem, there is no simple, general algorithm for constructing the solution for an equation having continuous, variable coefficients. Simple constructive statements about the solution are possible when the equation has *constant* coefficients.

Solutions for Homogeneous Equations with Constant Coefficients

Finding the solutions for an Nth–order homogeneous linear differential equation with constant coefficients can be reduced to the problem of finding the roots of a polynomial of degree N.

CHARACTERISTIC EQUATION

If L, given by (4.3), has constant coefficients, then substituting the exponential function e^{rx} for y in L[y] leads to

$$L[e^{rx}] = (r^N + a_{N-1} r^{N-1} + ... + a_0) e^{rx} = P_N(r) e^{rx}.$$

Then e^{rx} is a solution of the homogeneous equation

$$L[y]=0 \tag{4.5}$$

if and only if r is a root of the Nth–degree polynomial equation

$$P_N(r) = r^N + a_{N-1}r^{N-1} + ... + a_0 = 0 \tag{4.6}$$

The equation (4.6) is called the *characteristic* or *auxiliary equation* for the homogeneous equation (4.5). The roots of equation (4.6) are called the *characteristic* or *auxiliary roots* for L. Since (4.6) is an Nth–degree polynomial equation, there are N (not necessarily distinct) roots. If (4.6) has *distinct* roots $r_1,...,r_M$ then $P_N(r)$ can be factored into a product of the form

$$P_N(r) = (r-r_1)^a (r-r_2)^b ... (r-r_M)^m$$

with

$$a + b + .. + m = N.$$

We say that the roots $r_1, r_2, ..., r_M$ have *multiplicities* a,b,...,m respectively. Knowledge of the characteristic roots of L and their multiplicities provides complete information about the solutions of the homogeneous equation (4.5).

Theorem 4.4 Solutions of the Constant Coefficient Homogeneous Equation

Theorem 4.4 Suppose that $r = \mu$ is a root of multiplicity m for (4.6).
(a) if μ is real, then the m functions,

$$y_1(x) = e^{\mu x},$$

$$y_2(x) = xe^{\mu x}, ...$$

$$y_m(x) = x^{m-1}e^{\mu x}$$

are linearly independent solutions of (4.5).

(b) if μ is complex, with $\mu = \alpha + i\beta$, then $\alpha - i\beta$ is also a root for (4.6) and the 2m functions

$$u_1(x) = e^{\alpha x}\text{Cos } \beta x, \qquad v_1(x) = e^{\alpha x}\text{Sin } \beta x,$$

$$u_2(x) = xe^{\alpha x}\text{Cos } \beta x, \qquad v_2(x) = xe^{\alpha x}\text{Sin } \beta x,$$

$$... \qquad ...$$

$$u_m(x) = x^{m-1}e^{\alpha x}\text{Cos } \beta x, \qquad v_m(x) = x^{m-1}e^{\alpha x}\text{Sin } \beta x,$$

are linearly independent solutions of (4.5).

PARTICULAR SOLUTIONS

Theorem 4.2 implies that in order to construct the general solution for the inhomogeneous linear equation (4.2), it is sufficient to know N independent homogeneous solutions and *any* single solution to the inhomogeneous equation. We refer to a solution of the inhomogeneous equation as a *particular solution.*

Method of Undetermined Coefficients

In the special case that the coefficients a_k, $k=0,1,...,N-1$ are constants, it may be possible to find a particular solution for (4.2) by the method of *undetermined coefficients*. This method is practical if the forcing function f(x) is sufficiently simple, say a polynomial, an exponential, a sine or a cosine, or some simple combination of these. If the coefficients in the differential equation are variable or if f(x) is not simple, then this method is usually not workable.

Example 4.2 Undetermined Coefficients

4.2(A)

To find a particular solution for,

$$L[y(x)] = y''(x) + 3y'(x) + 2y(x) = 4x^2 + 2$$

note that $f(x) = 4x^2+2$ is a polynomial in x. Since derivatives of powers of x are (lower) powers of x, we are going to suppose that the particular solution is a polynomial in x. Moreover, we are going to suppose that the degree of this polynomial is 2, the same as the degree of f(x). We shall explain our reasons for this choice of degree in a moment.

If we suppose the particular solution is given by

$$y_P(x) = Ax^2 + Bx + C,$$

for unknown coefficients A, B and C then

$$L[y_P(x)] = 2A + 3(2Ax+B) + 2(Ax^2+Bx+C)$$
$$= 4x^2 + 2.$$

Equating coefficients of like powers of x on the two sides of this equation leads to

$$x^2: 2A \qquad = 4$$

$$x: 6A + 2B \qquad = 0$$

$$1: 2A + 3B + 2C = 2.$$

Then $A = 2$, $B = -6$, $C = 8$; i.e., $y_P(x) = 2x^2 - 6x + 8$.

Note that if we had chosen $y_P(x)$ to be a polynomial of degree *less* than the degree of f(x), then the degree of $L[y_P(x)]$ would be less than the degree of f(x); equating coefficients of like powers of x would lead to an inconsistent set of equations. If we chose $y_P(x)$ to be a polynomial of degree *higher* than the degree of f(x), then we would find that the coefficients of the terms of higher degree all turn out to be zero. Finally, we might wonder whether there is some other choice of $y_P(x)$ that would satisfy L[y(x)]=f(x). But Theorem 4.2 implies that it is sufficient to find *one* particular solution in order to form the general solution of the inhomogeneous equation.

4.2(B)

To find a particular solution for

$$L[y(x)] = y''(x) + 3y'(x) + 2y(x) = 3\text{Sin } 2x$$

we recall that derivatives of Sin ax involve Sin ax or Cos ax. Since f(x)= 3Sin 2x we suppose

$$y_P(x) = A\text{Sin } 2x + B\text{Cos } 2x,$$

for unknown coefficients A and B. Then

$$L[y_P(x)] = (-2A-6B)\text{Sin } 2x + (6A-2B)\text{Cos } 2x$$

and

$$L[y_P(x)] = 3\text{Sin } 2x \text{ for all x if and only if}$$

$$-2A - 6B = 3$$

$$6A - 2B = 0;$$

i.e., A = –3/20, B = –9/20, and

$$y_P(x) = (-3\text{Sin } 2x - 9\text{Cos } 2x)/20.$$

It may appear that in order for the method of undetermined coefficients to be effective, the user must be very good at guessing the form of the particular solution. Actually the form of the particular solution is dictated by the form of the forcing function. When f(x) is sufficiently simple, guesses may be formulated according to the following rules.

Theorem 4.5 The Method of Undetermined Coefficients

Theorem 4.5 Consider the inhomogeneous equation (4.2) in the case that L[y], given by (4.3) has constant coefficients.

(a) if the forcing function f(x) is of the form

$$f(x) = P_M(x) = x^M + b_{M-1}x^{M-1} + ... + b_0,$$

then a particular solution, $y_P(x)$, is of the form

$$y_P(x) = x^Q R_M(x) = x^Q(A_M x^M + A_{M-1}x^{M-1} + ... + A_0),$$

where Q is the smallest nonnegative integer for which no factor of $y_P(x)$ is a solution of the homogeneous equation.

(b) if the forcing function f(x) is of the form

$$f(x) = e^{ax}P_M(x),$$

then a particular solution $y_P(x)$ is of the form

$$y_P(x) = x^Q R_M(x) e^{ax}.$$

(c) if the forcing function f(x) is of the form

$$f(x) = e^{ax}P_M(x)\text{Sin }\Omega x,$$

or

$$f(x) = e^{ax}P_M(x)\text{Cos }\Omega x,$$

then a particular solution $y_P(x)$ is of the form

$$y_P(x) = x^Q R_M(x) e^{ax}\text{Sin }\Omega x + x^Q Q_M(x) e^{ax}\text{Cos }\Omega x\ .$$

Additional examples illustrating the method of undetermined coefficients are found in the solved problems.

Variation of Parameters

When the coefficients in L[y(x)] are not constant, or if f(x) is not sufficiently simple, then the method of undetermined coefficients is not practical. In that case we may instead use the method of *variation of parameters*.

If linearly independent solutions $y_1(x), y_2(x), ..., y_N(x)$ have been found for the homogeneous equation of order N, L[y(x)]=0, then we suppose that the inhomogeneous equation (4.2) has a particular solution of the form,

$$y_P(x) = C_1(x)y_1(x) + C_2(x)y_2(x) + ... + C_N(x)y_N(x), \tag{4.7}$$

where the N unknown functions $C_1(x), C_2(x), ..., C_N(x)$ are to be found.

In order to determine these N functions, we need N independent linear equations. One of the N equations is the equation (4.2) that we are trying to solve. Since we will be satisfied with *any* particular solution of (4.2), the remaining N–1 equations can be imposed at our discretion. These conditions will be imposed so as to make it easier to compute the C_k's. In particular, if we impose conditions that eliminate the derivatives of the C_k's from the expression for $L[y_P(x)]$, then (4.2) may lead to a differential equation for the C_k's that is simpler than the differential equation for y(x). Note that

$$y_P'(x) = C_1(x)y_1'(x) + C_2(x)y_2'(x) + ... + C_N(x)y_N'(x),$$
$$+ C_1'(x)y_1(x) + C_2'(x)y_2(x) + ... + C_N'(x)y_N(x).$$

Then the expression for $y_P'(x)$ contains no derivatives of C_k's if we require the C_k's to satisfy,

$$C_1'(x)y_1(x) + C_2'(x)y_2(x) + \dots + C_N'(x)y_N(x) = 0.$$

In fact the expression for $y_P^{(n)}(x)$ contains no derivatives of the C_k's for $n=1,2,\dots,N-1$ if we require that

$$C_1'(x)y_1^{(n-1)}(x) + \dots + C_N'(x)y_N^{(n-1)}(x) = 0 \tag{4.8}$$

for $n=1,\dots,N-1$.

We now compute $y_P^{(n)}(x)$ from (4.7), making use of the N–1 conditions (4.8). Substituting this expression into (4.2) and using the fact that $L[y_k(x)]=0$ for $k=1,\dots,N$, leads to

$$C_1'(x)y_1^{(N-1)}(x) + \dots + C_N'(x)y_N^{(N-1)}(x) = f(x). \tag{4.9}$$

The N–1 equations (4.8) together with (4.9) are a system of N equations for the N unknowns $C_1'(x),\dots, C_N'(x)$. The determinant of this system is the Wronskian $W[y_1,\dots,y_N](x)$, which is never zero since the functions $y_1(x),\dots,y_N(x)$ are linearly independent solutions of $L[y(x)]=0$.

We may apply Cramer's rule to this system in order to find the functions $C_1'(x),\dots, C_N'(x)$; i.e.,

$$C_k'(x) = \frac{f(x)\, W_k(x)}{W(x)} \quad \text{for } k=1,\dots,N. \tag{4.10}$$

Here $W(x)$ denotes the Wronskian, $W[y_1,\dots,y_N](x)$, and $W_k(x)$ denotes the determinant obtained when the kth column of the Wronskian is replaced by the column [0,0,...,1]. Then (4.10) may be integrated to obtain the coefficients $C_k(x)$ which are then substituted into (4.7) to get $y_P(x)$.

Example 4.3 Variation of Parameters

To find a particular solution for the inhomogeneous equation,

$$y^{(3)}(x) - y'(x) = f(x)$$

in terms of the forcing term $f(x)$, note that the general solution to the homogeneous equation is given by

$$y_H(x) = C_1 + C_2e^x + C_3e^{-x}.$$

Then we seek a particular solution of the form,

$$y_P(x) = C_1(x) + C_2(x)e^x + C_3(x)e^{-x},$$

where $C_1(x)$, $C_2(x)$, $C_3(x)$ are subject to the conditions,

$$C_1'(x) + C_2'(x)\, e^x + C_3'(x)\, e^{-x} = 0$$

$$C_2'(x)\, e^x - C_3'(x)\, e^{-x} = 0$$

$$C_2'(x)\, e^x + C_3'(x)\, e^{-x} = f(x).$$

Then

$$W[1,e^x,e^{-x}] = \begin{vmatrix} 1 & e^x & e^{-x} \\ 0 & e^x & -e^{-x} \\ 0 & e^x & e^{-x} \end{vmatrix} = 2$$

and replacing (respectively) the first, second and, third columns of this determinant by the column [0,0,1], we find:

$$W_1(x) = -2, \qquad W_2(x) = e^{-x}, \qquad W_3(x) = e^x.$$

According to (4.10),

$$C_1'(x) = -f(x),$$

$$C_2'(x) = \tfrac{1}{2}e^{-x}f(x), \qquad C_3'(x) = \tfrac{1}{2}e^x f(x),$$

and hence

$$C_1(x) = -\int_a^x f(y)\,dy$$

$$C_2(x) = \tfrac{1}{2}\int_a^x f(y)e^{-y}\,dy$$

$$C_3(x) = \tfrac{1}{2}\int_a^x f(y)e^{y}\,dy$$

Since we are looking for *any* particular solution, we may choose any value that is convenient for the lower limit a in the antiderivatives for the C_k's. Particular solutions that correspond to different choices for this limit then differ by a solution to the homogeneous equation. Finally, according to (4.7),

$$y_P(x) = \int_a^x (-1 + \tfrac{1}{2}e^{x-y} + \tfrac{1}{2}e^{-x+y})f(y)\,dy.$$

Additional examples of the method of variation of parameters are found in the solved problems.

SOLVED PROBLEMS

Operator Notation

PROBLEM 4.1

For each of the following equations, find the associated operator L:

(a) $y^{(4)}(x) - (2/x^2)\,y''(x) + (1/x^4)\,y(x) = 0$

(b) $y^{(3)}(x) - 4\,y''(x) + 2\,y'(x) = 0$

(c) $y^{(N)}(x) + a_{N-1}(x)\, y^{(N-1)}(x) + \dots + a_0(x)\, y(x) = 0$
Then apply the operators to the functions:

$$y(x) = e^{rx} \text{ and } y(x) = x^r.$$

SOLUTION 4.1

We shall use D to denote d/dx. Then the operators associated with the equations listed above are as follows:

(a) $L = D^4 - (2/x^2)D^2 + (1/x^4)$
(b) $L = D^3 - 4D^2 + 2D$
(c) $L = D^N + a_{N-1}(x)D^{N-1} + \dots + a_0(x)$.

Applying the operators to the functions $y(x) = e^{rx}$ and $y(x) = x^r$ leads to:

(a) $L[e^{rx}] = (\, r^4 - (2/x^2)r^2 + (1/x^4))e^{rx}$
$L[x^r] = (r(r-1)(r-2)(r-3) - 2r(r-1) + 1)x^{r-4}$

(b) $L[e^{rx}] = (r^3 - 4r^2 + 2r)e^{rx}$
$L[x^r] = r(r-1)(r-2)x^{r-3} - 4r(r-1)x^{r-2} + 2rx^{r-1}$

(c) $L[e^{rx}] = (r^N + a_{N-1}(x)r^{N-1} + \dots + a_0(x))e^{rx}$
$L[x^r] = r(r-1)\dots(r-N+1)x^{r-N}$
$\quad + a_{N-1}(x)r(r-1)\dots(r-N+2)x^{r-N+1}$
$\quad + \dots + a_0(x)x^r$

Linearity

PROBLEM 4.2

Suppose u(x), v(x) are two smooth functions satisfying the homogeneous equation

$$L[y(x)] = 0,$$

for L given by (4.3).

(a) Then for arbitrary constants C_1 and C_2, show that

$$L[\, C_1u(x)+C_2v(x)\,] = C_1\, L[u(x)] + C_2\, L[v(x)] = 0.$$

(b) If w(x) is a smooth function that satisfies the inhomogeneous equation $L[y(x)] = f(x)$, then for arbitrary constants C_1 and C_2, show that

$$L[\, C_1u(x)+C_2v(x)+w(x)\,] = f(x).$$

SOLUTION 4.2

Note first that if we let a_N=1, then the definition (4.3) implies

$$\sum_{k=1}^{N} a_k(d/dx)^k[C_1u(x)+C_2v(x)] = L[C_1u(x)+C_2v(x)]\,.$$

But, for $k=1,...,N$

$$a_k(d/dx)^k[C_1u(x)+C_2v(x)] = a_k(d/dx)^k[C_1u(x)] + a_k(d/dx)^k[C_2v(x)]$$
$$= C_1a_ku^{(k)}(x) + C_2a_kv^{(k)}(x),$$

hence

$$\sum_{k=1}^{N} [C_1a_ku^{(k)}(x) + C_2a_kv^{(k)}(x)] = C_1L[u(x)] + C_2L[v(x)].$$

It follows that

$$L[C_1u(x)+C_2v(x)] = C_1L[u(x)] + C_2L[v(x)] \tag{1}$$

and since $L[u] = L[v] = 0$, we have $L[C_1u+C_2v] = 0$. This shows part (a). In fact, we have proved that the set of solutions to the homogeneous linear equation $L[y]=0$ is a *SUBSPACE* of the linear space of smooth functions.

In the same way, we can show part (b) by writing

$$L[C_1u(x)+C_2v(x)+w(x)] = C_1L[u(x)] + C_2L[v(x)] + L[w(x)]$$
$$= 0 + 0 + f(x) = f(x).$$

We have shown that to any solution of the inhomogeneous equation, we can add an arbitrary linear combination of solutions of the homogeneous equation and the result is still a solution of the inhomogeneous equation.

Linear Dependence and the Wronskian

PROBLEM 4.3

Show that the functions $y_1(x)=x^2$ and $y_2(x)=x|x|$ are linearly independent.

SOLUTION 4.3

Note that $y_1(x)$ is positive for all x while $y_2(x)$ is positive where $x>0$ and negative when $x<0$. However, the absolute value of $y_1(x)$ equals the absolute value of $y_2(x)$ for all x. Then in order to have

$$C_1y_1(x) + C_2y_2(x) = 0 \text{ for all } x,$$

we would have to choose $C_1=C_2$ when $x<0$ and choose $C_1= -C_2$ when $x>0$. But there must be *one* choice of constants that makes the expression vanish for *all* x. Since no such constants can be found, the functions are independent.

PROBLEM 4.4

Suppose the functions $y_1(x)$ and $y_2(x)$ are smooth functions such that $W[y_1,y_2](x)$ is different from zero for some value of x. Then show that $y_1(x)$ and $y_2(x)$ are linearly independent.

Show also that $y_1(x)$ and $y_2(x)$ are linearly dependent if and only if $W[y_1,y_2](x)$ is zero for all x.

SOLUTION 4.4

To prove the first statement, suppose $W[y_1,y_2](x_0)$ is different from zero and that for constants C_1 and C_2, we have

$$C_1y_1(x) + C_2y_2(x) = 0 \tag{1}$$

for all x. Then it follows by differentiating, that for all x,

$$C_1y_1'(x) + C_2y_2'(x) = 0. \tag{2}$$

In particular these equations hold at the point $x=x_0$ where the Wronskian $W[y_1,y_2](x)$ is different from zero. But the Wronskian is the determinant of this set of two equations for the two unknowns C_1 and C_2. Since $W[y_1,y_2](x_0)$ does not equal zero, it follows that the equation (1) is satisfied only for C_1 and C_2 both equal to zero. Then $y_1(x)$ and $y_2(x)$ are linearly independent. Thus nonvanishing of the Wronskian implies independence of the functions.

To prove the second statement, suppose the Wronskian $W[y_1,y_2](x)$ vanishes for all values of x. It follows that the homogeneous linear algebraic equations (1), (2) have nontrivial solutions; i.e., nonzero constants C_1 and C_2 exist such that (1) and (2) hold for all x. Then $y_1(x)$ and $y_2(x)$ are linearly dependent. Conversely, if $y_1(x)$ and $y_2(x)$ are linearly dependent, then there exist nonzero constants such that (1) and (2) are satisfied for all x. But this is possible only if $W[y_1,y_2](x)=0$ for all x. Thus $y_1(x)$ and $y_2(x)$ are dependent *if and only if* $W[y_1,y_2](x)$ is zero for all x.

PROBLEM 4.5

Suppose $y_1(x)$ and $y_2(x)$ each satisfy,

$$L[y(x)] = y''(x) + a_1(x)y'(x) + a_0(x)y(x) = 0, \tag{1}$$

where $a_0(x)$ and $a_1(x)$ are continous for all x. Then show that the Wronskian, $W[y_1,y_2](x)$, satisfies the first order equation

$$d/dx\, W(x) = -a_1(x)W(x). \tag{2}$$

Conclude that when $y_1(x)$ and $y_2(x)$ solve (1) then either $W[y_1,y_2](x)=0$ for *all* x or $W[y_1,y_2](x)$ equals zero for *no* value of x.

SOLUTION 4.5

Note that

$$W[y_1,y_2](x) = \begin{vmatrix} y_1(x) & y_2(x) \\ y_1'(x) & y_2'(x) \end{vmatrix} = y_1y_2' - y_1'y_2$$

Then, differentiating with respect to x,

$$\begin{aligned} d/dx\, W[y_1,y_2](x) &= y_1y_2'' + y_1'y_2' - y_1'y_2' - y_1''y_2 \\ &= y_1y_2'' - y_1''y_2 \end{aligned}$$

We can use the differential equation (1) to replace $y_1''(x)$ and $y_2''(x)$, leading to

$$\begin{aligned} d/dx\, W[y_1, y_2](x) &= -y_1(a_1y_2' + a_0y_2) + (a_1y_1' + a_0y_1)y_2 \\ &= a_1(x)(y_2y_1' - y_1y_2') \\ &= -a_1(x)W[y_1, y_2](x). \end{aligned}$$

This is (2). Now it follows from Theorem 2.1 that

$$W[y_1,y_2](x) = W[y_1,y_2](x_0)e^{-A(x)} \tag{3}$$

where

$$A(x) = \int_{x_0}^{x} a_1(t)\, dt,$$

and x_0 denotes an arbitrary, fixed point. If an x_0 exists such that $W[y_1,y_2](x_0)$ is different from zero, then it follows from (3) that $W[y_1,y_2](x)$ is different from zero for all values of x. The alternative is that $W[y_1,y_2](x_0)$ is zero for all x_0.

Note that $W[y_1,y_2](x)$ vanishes for all x precisely when y_1 and y_2 are linearly dependent (i.e., when $y_1 = Cy_2$).

PROBLEM 4.6

Show that the result of the previous exercise holds for general N; i.e., that

$$d/dx\, W[y_1,y_2,\ldots, y_N](x) = -a_{N-1}(x)W[y_1,y_2,\ldots, y_N](x).$$

Thus, if $y_1,y_2,\ldots,y_N$ are N solutions of $L[y(x)]=0$ for L given by (4.3), then either $W[y_1,y_2,\ldots, y_N](x)$ equals zero for all x, or $W[y_1,y_2,\ldots, y_N](x)$ equals zero for no x.

SOLUTION 4.6

Recall that

$$W[y_1,y_2,\ldots, y_N](x) = \begin{vmatrix} y_1(x) & y_2(x) & \cdots & y_N(x) \\ y_1'(x) & y_2'(x) & \cdots & y_N'(x) \\ \cdot & \cdot & & \cdot \\ \cdot & \cdot & & \cdot \\ \cdot & \cdot & & \cdot \\ y_1^{(N-1)} & y_2^{(N-1)} & \cdots & y_N^{(N-1)} \end{vmatrix}$$

The derivative of an N by N determinant is a sum of N separate N by N determinants. The first of the N determinants is the original determinant with the first row differentiated. The second determinant has the second row differentiated, and so on through the Nth determinant, which has the Nth row differentiated. The determinant formed by differentiating the first row of $W[y_1,y_2,\ldots, y_N](x)$ is equal to zero since the first and second rows of this determinant are identical. Similarly, the determinants formed by differentiating rows 2 through N–1 of the Wronskian all are equal to zero since they have two identical rows. However, the determinant formed by differentiating the last row of $W[y_1,y_2,\ldots, y_N](x)$ does not have a pair of identical rows. This determinant is equal to

$$\text{Det} = \begin{vmatrix} y_1(x) & y_2(x) & \cdots & y_N(x) \\ y_1'(x) & y_2'(x) & \cdots & y_N'(x) \\ \cdot & \cdot & & \cdot \\ \cdot & \cdot & & \cdot \\ \cdot & \cdot & & \cdot \\ y_1^{(N)} & y_2^{(N)} & \cdots & y_N^{(N)} \end{vmatrix}$$

That is,

$$d/dx\, W[y_1,y_2,\ldots, y_N](x) = \text{Det}.$$

Now each of the functions $y_k(x)$ satisfies $L[y(x)]=0$ and hence each of them satisfies

$$y^{(N)}(x) = -a_{N-1}\, y^{(N-1)}(x) - \ldots - a_0\, y(x)\,.$$

We use this expression to replace each of the Nth derivatives that appear in the bottom row of the determinant Det. This implies that Det can be written as the following sum of N separate N by N determinants (obtained by splitting up the terms in the bottom row of Det):

$$\text{Det} = \begin{vmatrix} y_1(x) & y_2(x) & \cdots & y_N(x) \\ y_1'(x) & y_2'(x) & \cdots & y_N'(x) \\ \cdot & \cdot & & \cdot \\ \cdot & \cdot & & \cdot \\ \cdot & \cdot & & \cdot \\ -a_{N-1}y_1^{(N-1)} & -a_{N-1}y_2^{(N-1)} & \cdots & -a_{N-1}y_N^{(N-1)} \end{vmatrix}$$

$$+ \begin{vmatrix} y_1(x) & y_2(x) & \cdots & y_N(x) \\ y_1'(x) & y_2'(x) & \cdots & y_N'(x) \\ \cdot & \cdot & & \cdot \\ \cdot & \cdot & & \cdot \\ \cdot & \cdot & & \cdot \\ -a_{N-2}y_1^{(N-2)} & -a_{N-2}y_2^{(N-2)} & \cdots & -a_{N-2}y_N^{(N-2)} \end{vmatrix}$$

$$+ \dots + \quad + \begin{vmatrix} y_1(x) & y_2(x) & \dots & y_N(x) \\ y_1'(x) & y_2'(x) & \dots & y_N'(x) \\ \cdot & \cdot & & \cdot \\ \cdot & \cdot & & \cdot \\ \cdot & \cdot & & \cdot \\ -a_0y_1 & -a_0y_2 & \dots & -a_0y_N \end{vmatrix}$$

Note that all of these determinants except the first have two rows with the property that one row is a multiple of the other. Then each of these determinants is zero. Thus, only the first determinant has N independent rows. But this determinant is equal to

$$\text{Det} = \begin{vmatrix} y_1(x) & y_2(x) & \dots & y_N(x) \\ y_1'(x) & y_2'(x) & \dots & y_N'(x) \\ \cdot & \cdot & & \cdot \\ \cdot & \cdot & & \cdot \\ \cdot & \cdot & & \cdot \\ -a_{N-1}y_1^{(N-1)} & -a_{N-1}y_2^{(N-1)} & \dots & -a_{N-1}y_N^{(N-1)} \end{vmatrix}$$

so

$$\text{Det} = -a_{N-1}W[y_1,y_2,\dots, y_N](x).$$

This shows that,

$$d/dx\, W[y_1,y_2,\dots, y_N](x) = -a_{N-1}(x)W[y_1,y_2,\dots, y_N](x).$$

Then it follows that

$$W[y_1,y_2,\dots, y_N](x) = W[y_1,y_2,\dots, y_N](x_0)\, e^{-A(x)}$$

where

$$A(x) = \int_{x_0}^{x} a_{N-1}(t)\, dt.$$

Since the exponential is never zero, $W[y_1,y_2,\dots, y_N](x)$ is equal to zero for all x, or else it equals zero for no x. The Wronskian vanishes for all x precisely when the functions $y_1,y_2,\dots, y_N$ are linearly dependent.

General Solutions

PROBLEM 4.7

Find the general solution to each of the following equations:

(a) $(D-1)(D-2)(D-3)y(x) = 0$
(b) $(D^4-1)y(x) = 0$
(c) $(D^4+1)y(x) = 0$
(d) $D^2(D^2-1)y(x) = 0$
(e) $(D^2+1)^4y(x) = 0$

SOLUTION 4.7

part(a) Here

$$(D-1)(D-2)(D-3)y(x) = (D^3 - 6D^2 + 11D - 6)y(x) = 0$$

is a brief way of writing,

$$y^{(3)}(x) - 6y''(x) + 11y'(x) - 6y(x) = 0.$$

It is clear that the associated auxiliary equation is

$$(r-1)(r-2)(r-3) = 0$$

with auxiliary roots $r = 1, 2, 3$. Then according to Theorems 4.2 and 4.4, the general solution of this homogeneous equation is given by

$$y(x) = C_1e^x + C_2e^{2x} + C_3e^{3x}$$

for arbitrary constants C_1, C_2, C_3.

part(b) The auxiliary equation in this case is,

$$r^4 - 1 = 0,$$

with roots

$$r = 1, -1, i, -i \qquad (\text{here } i^2 = -1).$$

Then, according to Theorems 4.2 and 4.4, the general solution is formed from a linear combination of the four independent homogeneous solutions e^x, e^{-x}, Sin x, and Cos x. That is,

$$y(x) = C_1e^x + C_2e^{-x} + C_3\text{Sin } x + C_4\text{Cos } x.$$

Since the functions Sinh x and Cosh x are linear combinations of e^x and e^{-x},

$$\text{Sinh } x = (e^x - e^{-x})/2$$

$$\text{Cosh } x = (e^x + e^{-x})/2,$$

we can also write y(x) in the following alternative form,

$$y(x) = C_5\text{Sinh } x + C_6\text{Cosh } x + C_3\text{Sin } x + C_4\text{Cos } x.$$

part(c) The differential equation $(D^4 + 1)y(x) = 0$ leads to the auxiliary equation

$$r^4+1 = 0.$$

The roots of this equation are the four distinct fourth roots of -1; i.e.,

$$-1 = \exp(i\pi + 2n\pi i) \qquad n = \text{integer}$$

$$(-1)^{1/4} = \exp(i\pi/4 + in\pi/2) \qquad n=1,2,3,4.$$

We may write these four roots as follows,

$$r_1 = \alpha + i\alpha \qquad \alpha = (2)^{1/2}/2$$

$$r_2 = -\alpha + i\alpha$$

$$r_3 = -\alpha - i\alpha$$

$$r_4 = \alpha - i\alpha.$$

Then, according to Theorem 4.4, four independent homogeneous solutions are

$$y_1(x) = \exp(\alpha x)\text{Sin } \alpha x$$

$$y_2(x) = \exp(-\alpha x)\text{Sin } \alpha x$$

$$y_3(x) = \exp(-\alpha x)\text{Cos } \alpha x$$

$$y_4(x) = \exp(\alpha x)\text{Cos } \alpha x.$$

It follows from Theorem 4.2 that the general solution may then be written in the form,

$$y(x) = \exp[\alpha x](C_1\text{Sin } \alpha x + C_2\text{Cos } \alpha x) + \exp[-\alpha x](C_3\text{Sin } \alpha x + C_4\text{Cos } \alpha x).$$

part(d) Here the differential equation leads to the auxiliary equation

$$r^2(r^2-1) = 0,$$

whose roots are $r = 0, 0, 1, -1$. According to Theorem 4.4, part (a), the repeated root $r = 0$ generates two independent homogeneous solutions:

$$y_1(x) = 1 \text{ and } y_2(x) = x.$$

The simple roots $r = 1$ and $r = -1$ contribute one solution each to the general solution which is then given by

$$y(x) = C_1 + C_2x + C_3e^x + C_4e^{-x}.$$

part(e) Here $(D^2+1)^4y(x) = 0$ is the following equation of order eight:

$$y^{(8)}(x) + 4\, y^{(6)}(x) + 6\, y^{(4)}(x) + 4\, y''(x) + y(x) = 0.$$

The auxiliary equation,

$$(r^2 + 1)^4 = 0,$$

has the complex roots, $r = i$ and $r = -i$, each with multiplicity four. Then Theorem 4.4, part(b) implies that eight independent solutions of the homogeneous equation are the functions

$$y_1(x) = \text{Sin } x \qquad y_2(x) = \text{Cos } x$$

$$y_3(x) = x\text{Sin } x \qquad y_4(x) = x\text{Cos } x$$

$$y_5(x) = x^2\text{Sin } x \qquad y_6(x) = x^2\text{Cos } x$$

$$y_7(x) = x^3\text{Sin } x \qquad y_8(x) = x^3\text{Cos } x.$$

The general solution of the differential equation is then a linear combination of these eight solutions,

$$y(x) = \sum_{k=1}^{8} C_k y_k(x).$$

Particular Solutions: Undetermined Coefficients

PROBLEM 4.8

Consider the inhomogeneous differential equation

$$L[y(x)] = x^M + b_{M-1}x^{M-1} + ... + b_0,$$

in the case that L[y] given by (4.3) has constant coefficients. Show that if zero is not a root of the auxiliary equation, then

$$y_P(x) = A_M x^M + A_{M-1}x^{M-1} + ... + A_0$$

is a particular solution for the differential equation. If zero is a root of multiplicity Q for the auxiliary equation, then show that

$$y_P(x) = x^Q(A_M x^M + A_{M-1}x^{M-1} + ... + A_0),$$

is a particular solution of the inhomogeneous equation.

SOLUTION 4.8

For purposes of illustration, we shall first consider the following simple case: suppose y(x) solves

$$L[y(x)] = x. \tag{1}$$

Then differentiating both sides of (1) two times with respect to x shows that y(x) also satisfies

$$D^2L[y(x)] = 0. \tag{2}$$

Note that if $P_N(r) = 0$ is the auxiliary equation associated with equation (1), then

$$r^2P_N(r) = 0$$

is the auxiliary equation corresponding to (2). Applying Theorem 4.4 in connection with equation (2), we see that if zero is not a root of $P_N(r) = 0$, then

$$y(x) = A + Bx + C_1y_1(x) + ... + C_Ny_N(x)$$

is a general solution for (2). Here $y_1(x),...,y_N(x)$ denote N linearly independent solutions of $L[y]=0$.

If zero is a root of multiplicity Q for $P_N(r)=0$, then a general solution of (2) is of the form,

$$y(x) = Ax^Q + Bx^{Q+1} + C_1y_1(x) + ... + C_Ny_N(x).$$

In either case, y(x) solves (1) and

$$C_1y_1(x) + ... + C_Ny_N(x)$$

is a general homogeneous solution for the equation. Hence

$$y(x) = x^Q(A + Bx)$$

is a particular solution for (1). We have shown that in the case of this very simple forcing term, we can find a particular solution for the inhomogeneous equation (1) by finding a *homogeneous* solution for the higher order equation (2).

More generally, if y(x) solves

$$L[y(x)] = x^M + b_{M-1}x^{M-1} + ... + b_0 \tag{3}$$

then y(x) also solves the higher order homogeneous equation

$$D^{M+1}L[y(x)] = 0.$$

By the same line of reasoning used in the simpler case, we find that a particular solution for (3) is given by

$$y_P(x) = x^Q(A_Mx^M + A_{M-1}x^{M-1} + ... + A_0).$$

This method of replacing the inhomogeneous equation by a homogeneous equation of higher order will succeed so long as it is easy to find this associated homogeneous equation and it is also easy to find the additional roots for the new auxiliary equation. In other words, it will work as long as the forcing term is sufficiently simple.

PROBLEM 4.9 PRINCIPLE OF SUPERPOSITION

Find a particular solution for

$$L[y(x)] = x + 3e^{2x}, \tag{1}$$

where

$$L[y(x)] = y^{(3)}(x) + y'(x) - y(x).$$

SOLUTION 4.9

Here we are going to use the *principle of superposition* to reduce the problem to two simpler subproblems.

subproblem(a) Suppose u(x) solves

$$u^{(3)}(x) + u'(x) - u(x) = x. \tag{2}$$

Then we assume that u(x)= Ax + B and substitute into (2),

$$L[u(x)] = A - (Ax + B) = x.$$

Equating coefficients of like powers of x leads to

$$-A = 1 \text{ and } (A-B) = 0.$$

Then A=–1, B=–1, and u(x) = –x – 1.

subproblem(b) Suppose v(x) solves

$$v^{(3)}(x) + v'(x) - v(x) = 3e^{2x}. \tag{3}$$

Here we assume $v(x) = Ae^{2x}$ and substitute into (3),

$$L[v(x)] = 8Ae^{2x} + 2Ae^{2x} - Ae^{2x} = 3e^{2x}.$$

We solve this equation for A to obtain, A = 1/3 and the solution $v(x) = e^{2x}/3$.

Now

$$L[u(x)] = x$$
$$L[v(x)] = 3e^{2x}$$

$$L[u(x)] + L[v(x)] = x + 3e^{2x}.$$

But L[u(x)] + L[v(x)] = L[u(x)+v(x)] hence

$$L[u(x)+v(x)] = x + 3e^{2x}.$$

Thus, y(x) = u(x) + v(x) solves (1).

PROBLEM 4.10 NONRESONANT PERIODIC FORCING

Find a particular solution for the equation

$$y''(x) + dy'(x) + k^2y(x) = p\text{Sin } \Omega x + q\text{Cos } \Omega x. \tag{1}$$

Here we suppose that d, k, p, q and Ω denote given constants with d>0.

SOLUTION 4.10

According to Theorem 4.5 (since neither Sin Ωx nor Cos Ωx is a homogeneous solution) the equation (1) has a particular solution of the form

$$y(x) = C_1\text{Sin } \Omega x + C_2\text{Cos } \Omega x.$$

Then

$$L[y(x)] = (k^2 - \Omega^2)(C_1 \text{Sin } \Omega x + C_2 \text{Cos } \Omega x) \\ + d\Omega(C_1 \text{Cos } \Omega x - C_2 \text{Sin } \Omega x),$$

and the equation (1) is satisfied for all x if and only if

$$(k^2 - \Omega^2)C_1 - d\Omega C_2 = p$$
$$d\Omega C_1 + (k^2 - \Omega^2)C_2 = q.$$

That is,

$$C_1 = \frac{p(k^2 - \Omega^2) + qd\Omega}{(k^2 - \Omega^2)^2 + (d\Omega)^2}, \qquad C_2 = \frac{q(k^2 - \Omega^2) - pd\Omega}{(k^2 - \Omega^2)^2 + (d\Omega)^2}$$

After some algebraic simplification, we find:

$$C_1^2 + C_2^2 = \frac{p^2 + q^2}{(k^2 - \Omega^2)^2 + (d\Omega)^2}$$

Note that if we let

$$p^2 + q^2 = A^2,$$

then

$$\begin{aligned} p\text{Sin } \Omega x + q\text{Cos } \Omega x &= A((p/A)\text{Sin } \Omega x + (q/A)\text{Cos } \Omega x) \\ &= A(\text{Cos } \theta \text{ Sin } \Omega x + \text{Sin } \theta \text{ Cos } \Omega x) \\ &= A\text{Sin}(\Omega x + \theta) \end{aligned}$$

where we have introduced the notation $\theta = \text{Arctan}(q/p)$; i.e., A here denotes the *amplitude* of the forcing function and θ denotes the *phase angle*. Now if we let B be defined by,

$$(k^2 - \Omega^2)^2 + (d\Omega)^2 = B^2,$$

then

$$C_1^2 + C_2^2 = A^2/B^2$$

and

$$\begin{aligned} C_1\text{Sin } \Omega x + C_2\text{Cos } \Omega x &= A/B((BC_1/A)\text{Sin } \Omega x + (BC_2/A)\text{Cos } \Omega x) \\ &= (A/B)\text{Sin}(\Omega x + \phi). \end{aligned}$$

Here $\phi = \text{Arctan}(C_2/C_1)$ denotes the phase angle of the response.
Thus,

$$L[(A/B)\text{Sin}(\Omega x + \phi)] = A\text{Sin}(\Omega x + \theta),$$

which is to say, a sinusiodal forcing function with amplitude A and phase angle θ produces a sinusoidal solution having amplitude A/B and phase angle ϕ. Note that the amplitude of the solution is greater than or less than that of the forcing function, according to whether B is less than or greater than 1. In any case the amplitude of the solution does not vary with x. Compare this with the solution of the next problem.

PROBLEM 4.11 RESONANCE

Find a particular solution for

$$y''(x) + \Omega^2 y(x) = p\text{Sin } \Omega x + q\text{Cos } \Omega x, \qquad (1)$$

noting that Sin Ωx and Cos Ωx each satisfy the homogeneous equation.

SOLUTION 4.11

Note that both the functions Sin Ωx and Cos Ωx are homogeneous solutions for (1) and hence no function of the form

$$A\text{ Sin } \Omega x + B\text{ Cos } \Omega x$$

can be a particular solution for the equation (1). However, it is easily checked that neither xSin Ωx nor xCos Ωx solves the homogeneous equation. Then according to Theorem 4.5 part(c), we assume a particular solution of the form

$$y_P(x) = x(A\text{Sin } \Omega x + B\text{Cos } \Omega x) \qquad (2)$$

and we compute

$$y_P''(x) + \Omega^2 y_P(x) = 2\Omega(A\text{Cos } \Omega x - B\text{Sin } \Omega x).$$

It follows at once that $y_P(x)$ given by (2) is a particular solution for (1) provided

$$A = q/2\Omega \text{ and } B = -p/2\Omega;$$

i.e.,

$$y_P(x) = x(q\text{Sin } \Omega x - p\text{Cos } \Omega x)/2\Omega.$$

Note that the forcing function has constant amplitude equal to $(p^2 + q^2)^{1/2}$ but the amplitude of the particular solution increases steadily with x. This is an example of the phenomenon of *resonance*. We shall discuss resonance in the next chapter.

PROBLEM 4.12

Find a particular solution for

$$L[y(x)] = 18e^{-x} \qquad (1)$$

where

$$L[y(x)] = y^{(3)}(x) + 3y''(x) + 3y'(x) + y(x)$$

SOLUTION 4.12

The characteristic equation for the homogeneous differential equation $L[y(x)] = 0$ has the root $r = -1$ with multiplicity three. Then according to Theorem 4.4, e^{-x}, xe^{-x}, and x^2e^{-x} are independent solutions of the homogeneous equation. Then Theorem 4.5 suggests a particular solution of the form

$$y_P(x) = Ax^3e^{-x}.$$

A straightforward computation of derivatives leads to the result

$$L[y_P(x)] = 6Ae^{-x}$$

and then $y_P(x)$ solves (1) if and only if $6A = 18$; i.e. if

$$y_P(x) = 3x^3e^{-x}.$$

Particular Solutions: Variation of Parameters

PROBLEM 4.13

Use the method of variation of parameters to find a particular solution for the equation

$$y''(x) + k^2y(x) = f(x), \tag{1}$$

where

$$f(x) = \begin{cases} 1 & \text{if } 0 < x < 1 \\ 0 & \text{otherwise}. \end{cases} \tag{2}$$

SOLUTION 4.13

Note that the method of undetermined coefficients is impractical for f(x) given by (2). Instead, since the solution of the homogeneous equation is given by

$$y_H(x) = C_1\text{Sin } \Omega x + C_2\text{Cos } \Omega x,$$

for arbitrary constants, C_1 and C_2, we suppose there is a particular solution of the form,

$$y_P(x) = C_1(x)\text{Sin } \Omega x + C_2(x)\text{Cos } \Omega x. \tag{3}$$

The functions $C_1(x)$ and $C_2(x)$ are required to satisfy

$$C_1'(x)\text{Sin } \Omega x + C_2'(x)\text{Cos } \Omega x = 0$$

$$kC_1'(x)\text{Cos } \Omega x - kC_2'(x)\text{Sin } \Omega x = f(x).$$

Then

$$C_1'(x) = (1/k)f(x)\text{Cos } kx$$

$$C_2'(x) = (-1/k)f(x)\text{Sin } kx$$

and it follows by integration that

$$C_1(x) = 1/k \int_0^x f(y)\text{Cos } ky \, dy$$

$$C_2(x) = -1/k \int_0^x f(y)\text{Sin } ky \, dy.$$

Since $f(x)$ vanishes for $x < 0$, we may as well take the lower limit on these integrals to be zero.

Using these results in (3) leads to

$$y_P(x) = 1/k \int_0^x f(y)(\text{Cos } ky \text{ Sin } kx - \text{Sin } ky \text{ Cos } kx) \, dy$$

$$= 1/k \int_0^x f(y)\text{Sin } k(x-y) \, dy. \tag{4}$$

Note that in (4) we have the particular solution expressed in terms of the forcing function $f(x)$. For $f(x)$ given by (2), this results in

$$y_P(x) = 0 \text{ if } x < 0$$

and

$$y_P(x) = \begin{cases} 1/k \int_0^x \text{Sin } k(x-y) \, dy & \text{if } 0 < x < 1 \\ 1/k \int_0^1 \text{Sin } k(x-y) \, dy & \text{if } x > 1 \end{cases}$$

That is,

$$y_P(x) = \begin{cases} 0 & \text{if } x < 0 \\ (1-\text{Cos } kx)/k^2 & \text{if } 0 < x < 1, \\ (\text{Cos } k(x-1) - \text{Cos } kx)/k^2 & \text{if } x > 1\,. \end{cases}$$

The solution $y_P(x)$ is then "piecewise defined"; it is zero until $f(x)$ "turns on" at $x=0$, it varies periodically with x while $f(x)$ is "on", and then changes to a different periodic function after $f(x)$ "turns off" at $x=1$.

PROBLEM 4.14

Find a particular solution for the fourth order equation,

$$y^{(4)}(x) + a^4 y(x) = f(x), \tag{1}$$

where

$$f(x) = \begin{cases} 1 & \text{if } -1 < x < 1 \\ 0 & \text{otherwise .} \end{cases} \tag{2}$$

SOLUTION 4.14

As in the previous problem, the form of the function f(x) indicates the use of the method of variation of parameters. The solution to the homogeneous version of (1) is given by

$$y_H(x) = C_1 \text{Sin } ax + C_2 \text{Cos } ax + C_3 \text{Sinh } ax + C_4 \text{Cosh } ax,$$

and so we suppose there is a particular solution of the form

$$y_P(x) = C_1(x)\text{Sin } ax + C_2(x)\text{Cos } ax + C_3(x)\text{Sinh } ax + C_4(x)\text{Cosh } ax. \tag{3}$$

The coefficients $C_1(x), \dots, C_4(x)$ are required to satisfy

$$C_1'\text{Sin } ax + C_2'\text{Cos } ax + C_3'\text{Sinh } ax + C_4'\text{Cosh } ax = 0$$

$$C_1'\text{Cos } ax - C_2'\text{Sin } ax + C_3'\text{Cosh } ax + C_4'\text{Sinh } ax = 0$$

$$-C_1'\text{Sin } ax - C_2'\text{Cos } ax + C_3'\text{Sinh } ax + C_4'\text{Cosh } ax = 0$$

$$-C_1'\text{Cos } ax - C_2'\text{Sin } ax + C_3'\text{Cosh } ax + C_4'\text{Sinh } ax = f(x)/a^3$$

After some algebraic manipulation, we find that

$$2a^3C_1'(x) = -f(x)\text{Cos } ax,$$

$$2a^3C_2'(x) = -f(x)\text{Sin } ax,$$

$$2a^3C_3'(x) = f(x)\text{Cosh } ax,$$

$$2a^3C_4'(x) = f(x)\text{Sinh } ax,$$

which leads then to the result:

$$y_P(x) = (1/2a^3)\int_{-1}^{x} (\text{Sinh } ax \text{ Cosh } ay + \text{Cosh } ax \text{ Sinh } ax)f(y)dy$$

$$- (1/2a^3)\int_{-1}^{x} (\text{Sin } ax \text{ Cos } ay + \text{Cos } ax \text{ Sin } ax)f(y)dy$$

That is,

$$y_P(x) = (1/2a^3)\int_{-1}^{x} (\text{Sinh } a(x+y) - \text{Sin } a(x+y))f(y)\, dy \tag{4}$$

Equation (4) describes the particular solution in terms of the forcing function f(x). For f(x) given by (2) this reduces to

$$y_P(x) = \begin{cases} 0 & \text{if } x < -1 \\ (1/2a^3)\int_{-1}^{x} (\text{Sinh } a(x+y) - \text{Sin } a(x+y))dy & \text{if } -1<x<1 \\ (1/2a^3)\int_{-1}^{1} (\text{Sinh } a(x+y) - \text{Sin } a(x+y))dy & \text{if } 1<x \end{cases}$$

That is,

$$2a^4 y_P(x) = \begin{cases} 0 & \text{if } x < -1 \\ \text{Cosh } 2ax - \text{Cosh } a(x-1) & \\ \quad + \text{Cos } 2ax - \text{Cos } a(x-1) & \text{if } -1<x<1 \\ \text{Cosh } a(x+1) - \text{Cosh } a(x-1) & \\ \quad + \text{Cos } a(x+1) - \text{Cos } a(x-1) & \text{if } x>1 \end{cases}$$

As in the previous example, the piecewise nature of the forcing function produces a solution that is piecewise defined.

PROBLEM 4.15 AN INHOMOGENEOUS INITIAL VALUE PROBLEM

Use variation of parameters to find a particular solution for the equation

$$y''(x) - 2\,y'(x) + y(x) = f(x). \tag{1}$$

Find the unique particular solution that satisfies the initial conditions

$$y(0) = 0,\ y'(0) = 1. \tag{2}$$

SOLUTION 4.15

The auxiliary equation associated with (1) has the double root $r=1$, hence the homogeneous equation has independent solutions e^x and xe^x. Then (4.7) suggests a particular solution of the form

$$y_P(x) = C_1(x)e^x + C_2(x)xe^x, \tag{3}$$

where the coefficients $C_1(x)$ and $C_2(x)$ are required to satisfy

$$C_1'(x)e^x + C_2'(x)xe^x = 0$$

$$C_1'(x)e^x + C_2'(x)(x+1)e^x = f(x);\ \text{so}$$

$$C_1'(x) + C_2'(x)x = 0$$

$$C_1'(x) + C_2'(x)(x+1) = e^{-x}\,f(x).$$

Then

$$C_1'(x) = xe^{-x}f(x)$$
$$C_2'(x) = e^{-x}f(x),$$

and

$$C_1(x) = \int^x te^{-t}f(t)\,dt$$
$$C_2(x) = \int^x e^{-t}f(t)\,dt$$

Then it follows from (3) that

$$y_P(x) = \int^x (t+x)e^{x-t}f(t)\,dt.$$

In order to find a particular solution that satisfies the initial conditions (2), we choose a particular solution of the form

$$y_P(x) = \int_0^x (t + x)e^{x-t}f(t)\,dt + C_1e^x + C_2xe^x. \tag{4}$$

Note that,

$$y_P'(x) = \int_0^x (1 + t + x)e^{x-t}f(t)\,dt + 2xf(x) + C_1e^x + C_2(x+1)e^x.$$

Then

$$y_P(0) = C_1$$
$$y'(0) = C_1 + C_2$$

and the initial conditions (2) are satisfied if we choose C_1=0 and C_2=1. Substituting these choices into (4) then produces the desired particular solution.

In contrast to first order differential equations, which have a single solution, equations of order N have N independent solutions. To discuss this situation with precision requires the notion of dependence and independence which, in turn, requires the notion of the Wronskian of a set of functions. A set of smooth functions is dependent if their Wronskian *vanishes identically.*

Any homogeneous linear equation of order N has N linearly independent solutions. These solutions can then be used to form any solution of the homogeneous equation; in particular they can be used to form a solu

SUPPLEMENTARY PROBLEMS

tion that satisfies certain initial conditions. They may also be combined with a particular solution to form any solution for an inhomogeneous equation.

Finding N linearly independent solutions to a linear equation of order N having constant coefficients can be reduced to the problem of finding the roots of an N-th degree polynomial, solving the so called characteristic equation. *Once these roots are known, the solutions are easily found as described in Theorem 4.4.*

Finding particular solutions for inhomogeneous equations may be accomplished by either the method of undetermined coefficients *or the method of* variation of parameters. *The first method requires some educated guessing, while the second requires the knowledge of a full set of solutions for the associated homogeneous equation. As a result both methods are practically limited to equations with constant coefficients.*

For each of the following problems, write down the characteristic equation and the roots of the characteristic equation. Then give either the general solution or the indicated particular solution for the differential equation.

4.1 $y''(x) - 4y(x) = 0$

4.2 $3y''(x) + 2y'(x) = 0$

4.3 $y''(x) + 16y(x) = 0$

4.4 $y''(x) - 4y'(x) + 5y(x) = 0$

4.5 $y''(x) + y'(x) - 6y(x) = 0 \quad y(0) = 1 \quad y'(0) = 0$

4.6 $y''(x) + 4y(x) = \text{Cos } x$

4.7 $y''(x) + 9y(x) = \text{Sin } 3x$

4.8 $y''(x) - 4y'(x) + 5y(x) = 3e^{-x} + 2x^2$

4.9 $(D^3-8)y(x) = 0$

4.10 $(D^3-5D^2+6D)y(x) = 0$

4.11 $y''(x) - y'(x) + y(x) = 0, \quad y(0)=1, \quad y'(0) = 7/2.$

4.12 $y''(x) - 2y'(x) - 3y(x) = 0 \quad y(0)=0, \quad y'(0)=1.$

4.13 $y''(x) + 3y'(x) + 2y(x) = 0 \quad y(0)=0, \quad y'(0)=1$

4.14 $y''(x) - 3y'(x) - 10y(x) = 0 \quad y(0)=3, \quad y'(0)=15$

4.15 $(D^3-3D^2+D+3)y(x) = 0 \quad y(0)=0, \quad y'(0)=8, \quad y''(0)=24.$

ANSWERS TO SUPPLEMENTARY PROBLEMS

Find a particular solution for each of the following:

4.16 $y''(x) + a^2y(x) = k \text{ Cos } ax$ $\quad y(0) = y'(0) = 0.$

4.17 $6y''(x) + 5y'(x) - 6y(x) = x$

4.18 $y''(x) + 9y(x) = x^2e^{3x}$

4.19 $y''(x) - y'(x) - 2y(x) = x^2 + \text{Cos } x$

4.20 $y''(x) + 4y(x) = \text{Sin } 2x$

4.21 $y''(x) - 4y'(x) + 3y(x) = e^{2x}$ $\quad y(0) = y'(0) = 0.$

4.22 $y''(x) + y'(x) - 2y(x) = 3 \text{ Cos } 3x - 11 \text{ Sin } 3x$

4.23 $y''(x) + 5y'(x) + 6y(x) = x^2 + 2x$

4.24 $y''(x) - 2y'(x) + y(x) = e^x$

4.25 $y''(x) - 2y'(x) = 2$ $\quad y(0) = y'(0) = 0.$

4.1 $r^2-4 = 0$; $\quad r_{1,2} = -2,2$; $\quad y(x) = A\, e^{2x} + B\, e^{-2x}$.

4.2 $3r^2+2r = 0$; $\quad r_{1,2} = -2/3, 0$; $\quad y(x) = A + B\, e^{-2x/3}$.

4.3 $r^2+16 = 0$; $\quad r_{1,2} = -4i, 4i$; $\quad y(x) = A \text{ Cos } 4x + B \text{ Sin } 4x$.

4.4 $r^2-4r+5 = 0$; $\quad r_{1,2} = 2-i, 2+i$; $\quad y(x) = e^{2x}[A \text{ Cos } x + B \text{ Sin } x]$

4.5 $r^2+r-6 = 0$; $\quad r_{1,2} = -3,2$; $\quad y(x) = (3e^{2x} + 2e^{-3x})/5$.

4.6 $r^2+4 = 0$; $\quad r_{1,2} = -2i,2i$;
$y(x) = A \text{ Cos } 2x + B \text{ Sin } 2x + 1/3 \text{ Cos } x$.

4.7 $r^2+9 = 0$; $\quad r_{1,2} = -3i,3i$;
$y(x) = A \text{ Cos } 3x + B \text{ Sin } 3x - 1/6\, x \text{ Cos } 3x$.

4.8 $r^2-4r+5 = 0$; $\quad r_{1,2} = 2-i,2+i$;
$y(x) = e^{2x}[\, A \text{ Cos } x + B \text{ Sin } x\,] + 3/10\, e^{-x} + (44 + 80x + 50x^2)$

4.9 $r^3-8 = 0$; $\quad r = 2, -1 + i3^{1/2}, -1 - i3^{1/2}$;
$y(x) = Ae^{2x} + e^{-x}(\, B \text{ Cos } x3^{1/2} + C \text{ Sin } x3^{1/2}\,)$

4.10 $r^3-5r^2+ 6r = 0$; $r = 0,2,3$; $\quad y(x) = A + Be^{2x} + Ce^{3x}$.

4.11 $r^2-r+1 = 0$; $\quad r = [1 \pm i3^{1/2}]/2$
$y(x) = e^{x/2}[\, \text{Cos}(3^{1/2}x/2) + 2\, 3^{1/2} \text{ Sin}(3^{1/2}x/2)]$

4.12 $r^2-2r-3 = 0$; $\quad r = -1,3$; $\quad y(x) = (e^{3x}+2)/3 - e^{-x}-x$.

4.13 $r^2+3r+2 = 0$; $r = -1,-2$; $y(x) = e^{-x} - e^{-2x}$.

4.14 $r^2-3r-10 = 0$; $r = -2,5$; $y(x) = 3\,e^{5x}$.

4.15 $r^3-3r^2+r+3 = 0$; $r = -1,1,3$; $y(x) = 3e^{3x} - 2e^{x} - e^{-x}$.

4.16 $y(x) = (kx/2a)\ \text{Sin}\ ax$

4.17 $y(x) = A\,e^{2x/3} + b\,e^{-3x/2} - x/6 - 5/36$

4.18 $y(x) = e^{3x}(1-6x+9x^2)/162$

4.19 $y(x) = -(3\ \text{Cos}\ x + \text{Sin}\ x)/10 - (2x^2-2x+3)/4$

4.20 $y(x) = -(x/4)\ \text{Cos}\ 2x$

4.21 $y(x) = (e^{x} - 2e^{2x} + e^{3x})/2$

4.22 $y(x) = e^{x} - e^{-2x} + \text{Sin}\ 3x$

4.23 $y(x) = (18x^2 + 6x - 11)/108$

4.24 $y(x) = A\,e^{x} + B\,xe^{x} + \frac{1}{2}x^2e^{x}$

4.25 $y(x) = (e^{2x}-1)/2 - x$

5

Applications of Higher Order Equations

The previous chapter presented some fundamental ideas associated with solving linear differential equations of order higher than one. In particular, we found that the general solution of a homogeneous linear equation of order N consists of a linear combination of N linearly independent solutions. In addition, we saw how to construct general solutions for Nth-order homogeneous equations and particular solutions for inhomogeneous equations of order N if the equations have constant coefficients.

In this chapter we shall describe a number of physical applications that lead to differential equations of order higher than one. In particular we shall see that a surpising number of physical systems can be modeled by a second order differential equation of the form

$$Ay''(t) + By'(t) + Cy(t) = F(t).$$

Systems that are modeled by this equation include electrical circuits and mechanical vibrations of various types. In a limited number of examples, an equation of order higher than two is required in order to model the behavior of the physical system.

TERMINOLOGY FOR APPLICATIONS

Before considering examples of specific physical systems we establish the terminology we shall use in discussing systems modeled by the second order equation above.

FORCED AND UNFORCED RESPONSE

We shall consider examples of *unforced* or *homogeneous* response in which the forcing function F(t) is zero. Then the character of the response is largely determined by the coefficient, B, of the first-order term. The first order term can be shown to be related to energy dissipation in the physical system.

TRANSIENT AND STEADY STATE SOLUTIONS

We shall also consider *forced response* in which F(t) is not zero. Then the solution will be shown to consist of a *transient component* and a *steady state component.* The transient component of the solution can often be interpreted as the response of the physical system to the initial state while the steady state component can be viewed as the response to the forcing.

RESONANCE

When the forcing function is periodic, the character of the response may vary with the forcing frequency. In particular, at certain values of the forcing frequency we encounter the phenomenon of resonance in which the magnitude of the output may be many times larger than the magnitude of the input.

AMPLITUDE AND PHASE ANGLE

In this chapter we will very often encounter solutions of the form,

$$x(t) = p\text{Cos}\Omega t + q\ \text{Sin}\Omega t.$$

This solution can always be written in the alternative forms

$$x(t)=A\ \text{Cos}(\Omega t-\varphi),\ \text{or}\ x(t)=A\ \text{Sin}(\Omega t+\psi)$$

where $A=\sqrt{p^2 + q^2}$ and $\varphi=\text{Arctan}(q/p)$ and $\psi=\text{Arctan}(p/q)$. Then A is called the *amplitude* of the solution and φ or ψ is called the *phase angle.* To see the equivalence of these expressions note that

$$\text{Cos}(\Omega t-\varphi)= \text{Cos}\Omega t\ \text{Cos}\varphi + \text{Sin}\Omega t\ \text{Sin}\varphi$$

$$\text{Sin}(\Omega t+\psi) = \text{Sin}\Omega t\ \text{Cos}\psi + \text{Cos}\Omega t\ \text{Sin}\psi$$

Then, since

$$x(t) = A((p/A)\cos\Omega t + (q/A)\sin\Omega t)$$

it is clear that the expressions are identical if φ and ψ are chosen such that $p/A=\cos\varphi=\sin\psi$ and $q/A=\sin\varphi=\cos\psi$; i.e., if

$$\varphi=\arctan(q/p), \text{ and } \psi=\arctan(p/q).$$

MATHEMATICAL MODELS FOR PHYSICAL SYSTEMS

Mechanical Vibrations—Spring-mass Systems

Consider a body of mass m attached to one end of an elastic spring whose other end is held fixed. If the mass is given an initial displacement and is then released from rest, the elastic restoring force of the spring causes the mass to execute a periodic "bobbing" motion. If the mass and spring are submerged in a viscous fluid, then, in addition to the elastic spring force, the mass is affected by a viscous damping force. Finally, we may wish to consider the additional effect of an external force applied to the mass. Then according to Newton's second law, the sum of the forces acting on the mass must equal the rate of change of the momentum of the mass. In order to make this statement precise, we must introduce a coordinate system.

Suppose the motion of the mass on the spring is 1-dimensional; i.e., the mass moves back and forth along a straight line. Then label the location of the mass when the spring is in its equilibrium, unstretched state as $x=0$ and denote the position of the mass as a function of time by $x(t)$. We shall take $x(t)$ to be positive when the spring is in a stretched state and negative when the spring is compressed. This system is pictured in Figure 5.1.

FORCED VIBRATION

The forces acting on the mass can be characterized as follows:

$$F_{spring} = \text{the spring force} = -Kx(t)$$

$$F_{friction} = \text{frictional force} = -Cx'(t)$$

$$F_{applied} = \text{the applied force} = F(t)\,.$$

The spring force is assumed to be proportional to the amount by which the spring is stretched. This is called *Hooke's law.* The positive constant of pro-

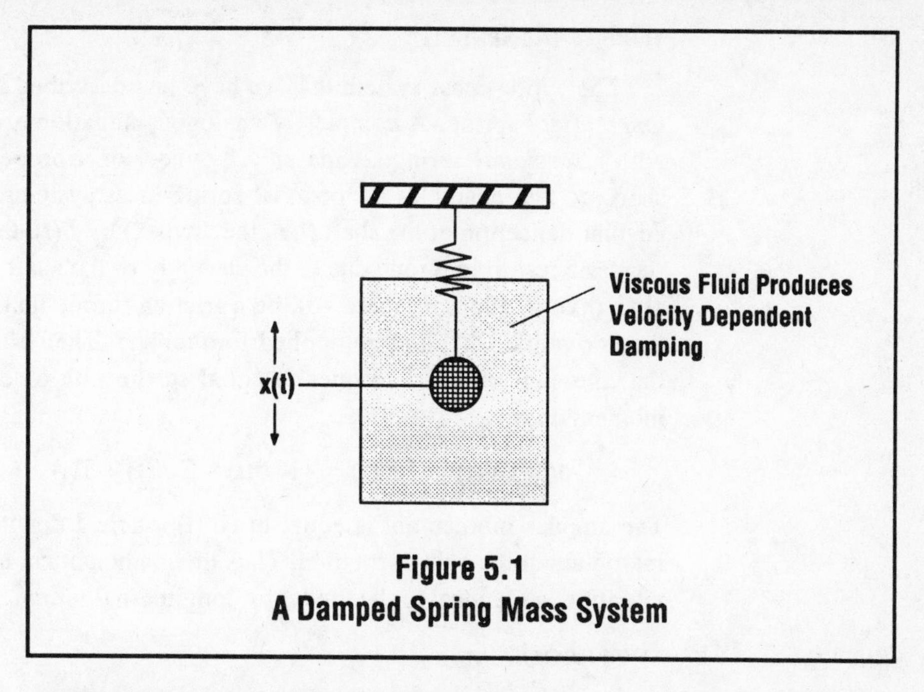

Figure 5.1
A Damped Spring Mass System

portionality is denoted by K and the presence of the minus sign in the expression for the spring force reflects the fact that when the spring is stretched the force is contractive and if the spring is compressed the force is expansive. The friction force is assumed here to be proportional to the velocity of the mass, i.e. to $x'(t)$. Here C is a positive constant and the minus sign in the expression indicates that the friction force is in the direction opposite to the direction of the velocity.

The momentum of the mass equals $mx'(t)$ hence Newton's law assumes the form

$$d/dt(mx'(t)) = mx''(t) = -Kx(t) - Cx'(t) + F(t) \tag{5.1}$$

FREE VIBRATION OF A SPRING MASS SYSTEM

Equation (5.1) is the governing equation for the so called "forced, damped spring mass system". When F(t) is zero, we say that (5.1) describes *free vibration* or *unforced vibration.* In this case, the nature of the motion executed by the mass is determined by the *damping parameter* C. There are four possible cases,

1. If $C=0$ then the motion is said to be *undamped.*
2. If $4\,Km > C^2 > 0$ then the motion is said to be *underdamped.*
3. If $C^2 = 4\,Km$ then the motion is said to be *critically damped.*
4. If $C^2 > 4K\,m$ then the motion is said to be *overdamped.*

TORSIONAL SPRING

The spring-mass system that we have just described is an example of a *longitudinal* spring. A completely analogous situation arises in connection with a *torsional* spring. A massive flywheel or a propellor on a rotating shaft are two examples of torsional spring-mass systems. If we denote the angular deflection of the shaft (i.e., the "twist") by $\vartheta(t)$, then the shaft experiences a restoring torque due to the elasticity of the shaft that is proportional to $\vartheta(t)$. In addition, there will be a friction torque that is proportional to the derivative $\vartheta'(t)$ and an applied torque $T(t)$. Then Newton's law states that the sum of these torques is equal to the rate of change of angular momentum; i.e.,

$$d/dt(I\vartheta'(t)) = I\vartheta''(t) = -K\vartheta(t) - C\vartheta'(t) + T(t)\,.$$

The angular momentum is equal to $I\vartheta'(t)$ where I denotes the moment of inertia about the axis of rotation. Thus the mathematical model for torsional vibration is identical to the model for longitudinal vibration.

THE PENDULUM

The weight on a pendulum discussed in exercise 3.10 is a third system that is very like the longitudinal and torsional spring systems. The angular displacement of the pendulum, $\vartheta(t)$, satisfies the nonlinear equation,

$$mL\vartheta''(t) = -mg\,\mathrm{Sin}\vartheta(t) - C\vartheta'(t).$$

However, when the amplitude of the pendulum is small, $\mathrm{Sin}\vartheta(t)$ is approximately equal to $\vartheta(t)$ and the nonlinear equation reduces to the following linear equation,

$$mL\vartheta''(t) = -mg\vartheta(t) - C\vartheta'(t).$$

Thus, for small amplitudes, the motion of the pendulum is governed by the same equation as the longitudinal and torsional springs.

Electrical Circuits

SIMPLE CIRCUITS

A simple electrical circuit is another example of a physical system whose behavior is governed by a linear second-order ordinary differential equation. A circuit consists of a closed loop through which an electric charge is driven by an imposed electromotive force. The charge is denoted by Q and is measured in coulombs while the electromotive force is denoted by E and is measured in volts. The movement of charge induced by the electromotive force produces a current denoted by I and related to Q by

$$I = dQ/dt. \tag{5.2}$$

As a result of experimental observation it has been determined that as the charge circulates through the conducting loop it experiences voltage drops of three types:

RESISTANCE LOSS—a drop in voltage proportional to the current I:

$$\Delta V_R = R\,I \tag{5.3}$$

The constant of proportionality R is called the *resistance*. It is measured in ohms.

CAPACITANCE LOSS—a drop in voltage proportional to the charge Q:

$$\Delta V_C = Q/C \tag{5.4}$$

The reciprocal of the constant of proportionality C is called the *capacitance* and is measured in farads.

INDUCTANCE LOSS—a drop in voltage proportional to the time derivative of the current, dI/dt

$$\Delta V_L = L dI/dt \tag{5.5}$$

The constant of proportionality L is called the *inductance* and is measured in henrys.

KIRCHHOFF'S VOLTAGE LAW

The flow of current in the circuit is governed by *Kirchhoff's voltage law* which states that the algebraic sum of the voltages in a closed circuit equals zero. That is,

$$E(t) - \Delta V_R - \Delta V_C - \Delta V_L = 0 \tag{5.6}$$

Substituting (5.3), (5.4), and (5.5) into (5.6) leads to the differential equation

$$L\, dI/dt + R\, I(t) + 1/C\, Q(t) = E(t).$$

According to (5.2), this equation is equivalent to either of the following equations:

$$L\, d^2I/dt^2 + R\, dI/dt + 1/C\, I(t) = dE/dt \tag{5.7}$$

$$L\, d^2Q/dt^2 + R\, dQ/dt + 1/C\, Q(t) = E(t) \tag{5.8}$$

Since the equations (5.7), (5.8) each involve just a single unknown function, when the imposed electromotive force E(t) is given then we can solve either (5.7) for I(t) or (5.8) for Q(t) provided initial conditions are specified for the unknown function and its first derivative. Then I(t) and Q(t) are related by (5.2).

Transverse Deflection of an Elastic Beam

The applications involving mechanical vibration and electrical circuits each led to differential equations of order two. In fact most physical systems that can be modeled by differential equations are described by equations of order one or two. In a few cases, however, an equation of order higher than two is required.

Consider an elastic beam with centerline lying along the x-axis subjected to a transverse load distributed along the length of the beam. Let $y(x)$ denote the transverse displacement of a point on the beam's centerline at position x along the beam. Then if $y(x)$ and $y'(x)$ are each small, it is a reasonable approximation of reality to say that $M(x)$, the internal bending moment acting at position x in the beam, is proportional to $y''(x)$, the curvature in the beam centerline at position x. More precisely,

$$M(x) = EIy''(x) \tag{5.9}$$

where E denotes a material dependent property known as *Young's modulus* and I denotes the moment of inertia of the beam cross section at x; i.e. I is a parameter depending on the shape of the beam cross section at the position x.

If $S(x)$ denotes the internal shear force acting at x, then

$$dM(x)=S(x)dx;$$

$$S(x)=M'(x)=EIy^{(3)}(x)\,. \tag{5.10}$$

Similarly, if $F(x)$ denotes the applied transverse load per unit length, then $dS(x)= F(x)dx$; i.e., $S'(x)=F(x)$. Combining these two results with (5.9) then leads to,

$$EIy^{(4)}(x) = F(x). \tag{5.11}$$

Equation (5.11) governs the (static) transverse deflection of an elastic beam subjected to a distributed transverse load given by $F(x)$. In the special case that the beam rests on an elastic foundation, then $F(x)=W(x)-Ky(x)$, where $W(x)$ denotes the applied loading (per unit length) and K denotes the elastic constant of the foundation. A railroad track on its bed is one example of a system that may be treated as a beam on an elastic foundation. The load $W(x)$ is applied by the train, and the track bed depresses but resists with a force that may be assumed to be proportional to the amount of depression. When the applied loading is known, then we may be able to solve (5.11) for $y(x)$.

SOLVED PROBLEMS

Free Response of Physical Systems

PROBLEM 5.1

Consider a mass m suspended on a longitudinal spring with spring constant equal to K. If the mass is given an initial displacement and released from rest, compute the response in each of the following cases:

(a) undamped $C=0$
(b) underdamped $C > 0$ but $C^2 < 4Km$
(c) critically damped $C^2 = 4Km$
(d) overdamped $C^2 > 4Km$.

SOLUTION 5.1

Let the displacement of the mass from the equilibrium position be denoted by x(t). Then x(t) satisfies equation (5.1) with $F(t)=0$; i.e.,

$$mx''(t) + Cx'(t) + Kx(t) = 0\,. \tag{1}$$

If we denote the amount of the initial displacement by a, then the initial conditions that the mass is displaced by an amount a and released from rest are

$$x(0) = a, \qquad x'(0) = 0. \tag{2}$$

The solution is given in four parts:

part (a) In the undamped case, $C=0$, the auxiliary equation associated with (1) is

$$mr^2 + K = 0,$$

with roots $r = i\omega, -i\omega$, where $\omega^2 = K/m$. Then the corresponding general solution for (1) is given by

$$x(t) = A\text{Cos } \omega t + B\text{Sin } \omega t, \tag{3}$$

for arbitrary constants A and B. Then, (3) implies that $x(0) = A$ and $x'(0) = \omega B$, and it follows that the initial conditions are satisfied if $A = a$, $B = 0$; hence

$$x(t) = a \text{ Cos } \omega t \tag{4}$$

is the undamped solution. This solution is periodic with frequency ω equal to $\sqrt{K/m}$; this frequency is referred to as the *natural frequency* of the undamped system. Note that ω is directly proportional to the square root of K and inversely proportional to the square root of m. So the system oscillates more rapidly if K is increased (i.e., the spring is made stiffer) and more slowly if the mass is increased.

part(b) In the underdamped case $C > 0$ but $C^2 < 4Km$, the auxiliary equation associated with (1) is

$$mr^2 + Cr + K = 0.$$

The roots of this equation are given by

$$r = (-C \pm \sqrt{(C^2 - 4Km)})/2m$$
$$= -\beta \pm i\omega_D$$

where

$$\beta = C/2m \text{ and } \omega^2_D = (4Km - C^2)/4m^2 = \omega^2 - \beta^2 .$$

The corresponding general solution of the differential equation (1) is then

$$x(t) = e^{-\beta t}(A\text{Cos } \omega_D t + B\text{Sin } \omega_D t). \quad (5)$$

According to (5), $x(0) = A$ and $x'(0) = -\beta A + \omega_D B$, and it follows that the initial conditions will be satisfied if we choose the arbitrary constants A and B such that: $A=a$ and $B= a\beta/\omega_D$. Then the solution of the initial value problem is

$$x(t) = a\, e^{-\beta t}(\text{Cos } \omega_D t + (\beta/\omega_D)\text{Sin } \omega_D t). \quad (6)$$

Note that the solution (6) reduces to (4) when C goes to zero. For C greater than zero but less than the critical value $C = \sqrt{4Km}$, the solution oscillates with steadily decreasing amplitude at the *damped natural frequency* $\omega_D = \sqrt{(\omega^2 - \beta^2)}$. (Figure 5.2).

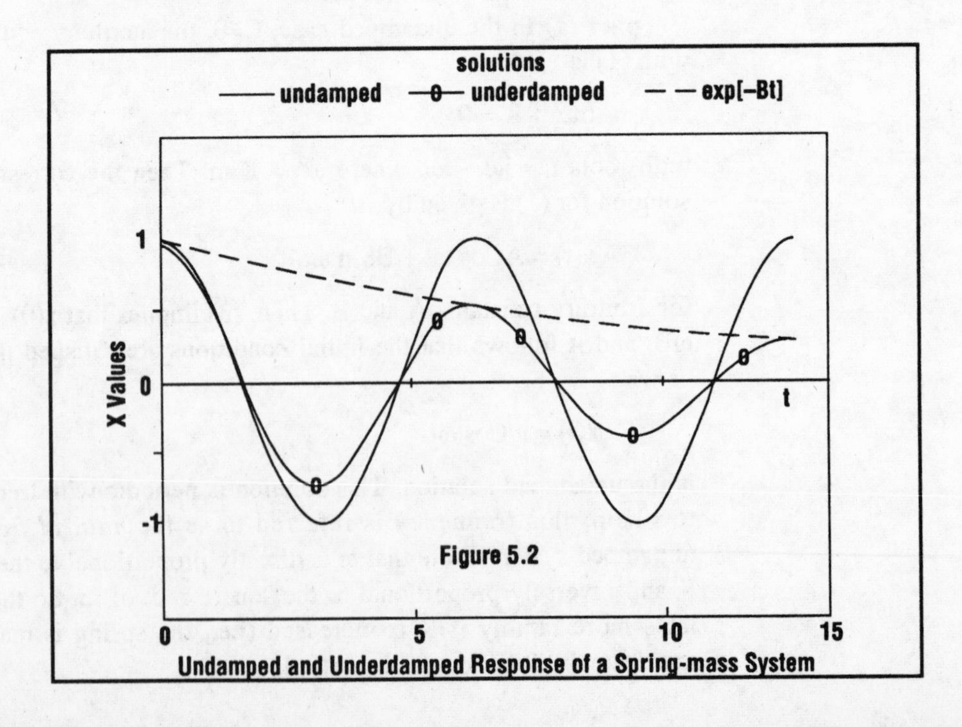

Figure 5.2

Undamped and Underdamped Response of a Spring-mass System

part (c) Note that ω_D is less than ω, the undamped natural frequency. As the damping increases from zero, the damped natural frequency continues to decrease until, finally, it reaches the value zero and no oscillation occurs. In the critically damped case $C^2 = 4Km$, the auxiliary equation has a double root at the value $r = \Omega = \sqrt{K/m}$. Then the general solution to the differential equation (1) is given by

$$x(t) = Ae^{-\omega t} + Bte^{-\omega t} \tag{7}$$

Then $x(0) = A$ and $x'(0) = -\omega A + B$, which leads to $A = a$, $B = a\omega$ and

$$x(t) = a(e^{-\omega t} + \omega te^{-\omega t}) \tag{8}$$

is then the solution to the initial value problem. While the underdamped solution given by (5) oscillates with exponentially decreasing amplitude, the critically damped solution given by (8) decreases to zero without oscillating. Thus, when the damping reaches the critical value $\sqrt{4Km}$, the damping is just sufficient to suppress oscillation. Thus the mass released from rest does not oscillate but merely returns slowly to the equilibrium position.

part (d) Increasing damping beyond the critical value so that C^2 is greater than $4Km$ produces the condition of overdamping. Here the auxiliary equation associated with (1) has distinct real roots given by $r_{1,2} = -\beta \pm \lambda$ where $\beta = C/2m$ and $\lambda = \sqrt{(\beta^2 - \omega^2)}$. The corresponding solution for (1) is then

$$x(t) = Ae^{(-b-\lambda)t} + Be^{(-b+\lambda)t} \tag{9}$$

and since $x(0) = A + B$ and $x'(0) = (-\beta-\lambda)A + (-\beta+\lambda)B$ it follows that the solution of the initial value problem is

$$x(t) = \frac{a}{2\lambda}((\lambda - \beta)\, e^{(-\beta-\lambda)t} + (\lambda + \beta)\, e^{(-\beta+\lambda)t}) \tag{10}$$

Note that by definition, β is greater than λ. Then $-\beta-\lambda < -\beta+\lambda < 0$ so both the exponents appearing in the solution $x(t)$ given by (10) are negative, and the solution decreases steadily to zero. Note further that the overdamped solution (10) decreases to zero less rapidly than the critically damped solution, (Figure 5.3). When the damping is increased beyond the critical value $\sqrt{4Km}$, the damping not only suppresses oscillation, it further impedes the motion of the mass released from rest so the return to the equilibrium position occurs more slowly.

Heavy doors are often equipped with spring devices that pull the door shut but the motion is sufficiently damped that the door does not slam. For this to work properly, the damping must equal or exceed the critical value.

PROBLEM 5.2

Consider the spring mass system of Problem 5.1 in the critically damped and overdamped cases if the initial conditions are changed to,

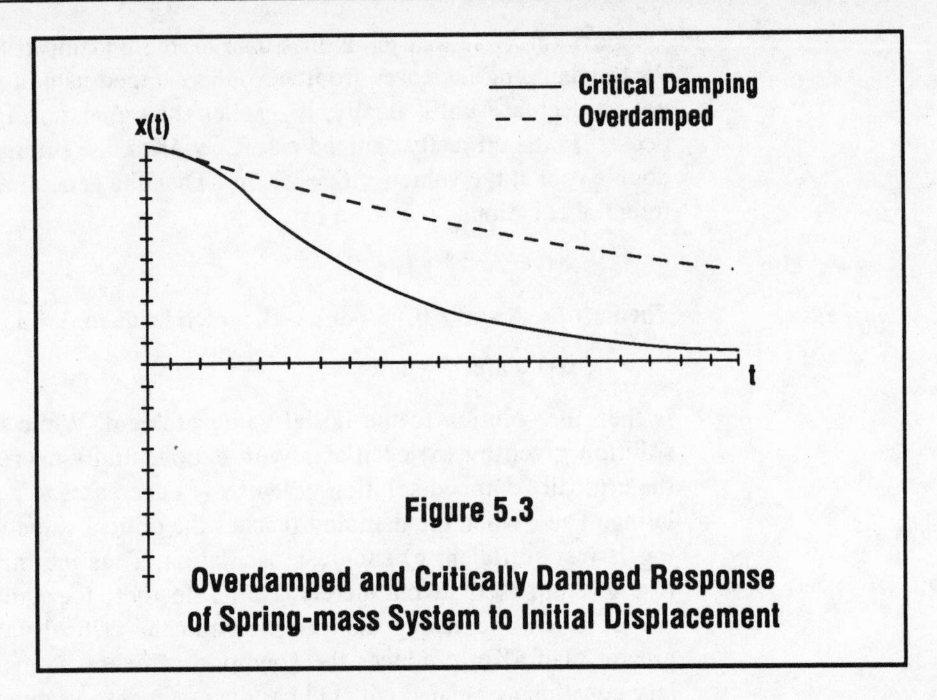

Figure 5.3

Overdamped and Critically Damped Response of Spring-mass System to Initial Displacement

$$x(0) = 0, \qquad x'(0) = b\,. \tag{1}$$

These initial conditions correspond to the physical situation in which the mass is given a sharp blow so that it starts from the rest position with a nonzero velocity. This initial value problem for the spring-mass system is a reasonable approximation of the operation of a shock absorber in an automobile; i.e., the shock absorber is essentially a damped spring-mass system subject to the initial conditions (1) when the car hits a bump.

SOLUTION 5.2

In the critically damped case we obtain the general solution

$$x(t) = Ae^{-\omega t} + Bte^{-\omega t}\,. \tag{2}$$

As in Problem 5.1, we compute $x(0) = A$, $x'(0) = -\omega A + B$. Then the initial conditions (1) lead to the result A=0 and B=b and the solution,

$$x(t) = bte^{-\omega t}\,. \tag{3}$$

In the overdamped case, the auxiliary equation has distinct, real roots and the general solution of the differential equation is

$$x(t) = Ae^{(-\beta-\lambda)t} + Be^{(-\beta+\lambda)t}\,. \tag{4}$$

Then $x(0) = A + B$ and $x'(0) = (-\beta-\lambda)A + (-\beta+\lambda)B$ and the initial conditions (1) lead to the solution,

$$x(t) = \frac{b}{2\lambda}\left(e^{(-\beta+\lambda)t} - e^{(-\beta-\lambda)t}\right). \tag{5}$$

Figure 5.4 shows the critically damped and overdamped solutions (3) and (5) for the initial conditions (1). Each increases to a maximum value and then decreases to zero. The critically damped solution rises to a greater maximum than the overdamped solution; thus, increasing the damping beyond the critical value needed to suppress oscillation further inhibits the motion of the mass. In both cases note that the spring in the shock absorber is compressed by the bump but that the damping suppresses any oscillation. Then the car does not bob up and down after hitting a bump.

PROBLEM 5.3

The energy in a system is defined to be the sum of the kinetic and potential energies. For the spring-mass system, the kinetic and potential energies are given by

Kinetic Energy = $mx'(t)^2/2$ (due to the motion of the mass)

Potential Energy = $Kx(t)^2/2$. (due to elastic spring force)

Then the total energy at time t equals

$$E(t) = 1/2\, m\, x'(t)^2 + 1/2\, K\, x(t)^2 . \tag{1}$$

Compute the energy for the spring-mass system of Problem 5.1 in the undamped and underdamped cases.

SOLUTION 5.3

Substituting the undamped solution from exercise 5.1 into (1), we have

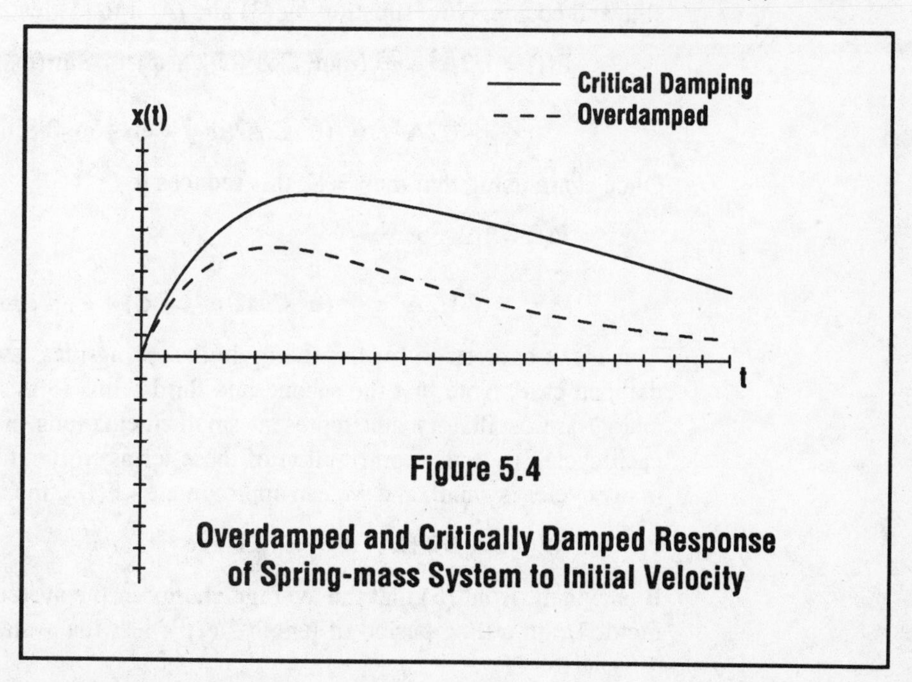

Figure 5.4

Overdamped and Critically Damped Response of Spring-mass System to Initial Velocity

$$E(t) = 1/2\, m(-a\omega \,\text{Sin}\, \omega t)^2 + 1/2\, K\,(a\text{Cos}\, \omega t)^2$$
$$= 1/2\, Ka^2 . \qquad (\text{ recall } m\omega^2 = K\,) \tag{2}$$

Thus, the total energy is constant when the damping is zero. A physical system in which the total energy is constant is said to be a *conservative system.* Thus, a spring-mass system with no damping is conservative. As we shall see, when there is damping in the system, it is no longer conservative.

The underdamped solution from Problem 5.1 can be written in the form,

$$x(t) = a\, \frac{\omega}{\omega_D}\, e^{-\beta t}\, (\frac{\omega_D}{\omega} \text{Cos}\, \omega_D t + \frac{\beta}{\omega} \text{Sin}\, \omega_D t\,)$$
$$= A\, e^{-\beta t}\, (\text{Sin}\, \varphi\, \text{Cos}\, \omega_D t + \text{Cos}\, \varphi\, \text{Sin}\, \omega_D t)$$
$$= A\, e^{-\beta t}\, \text{Sin}\, (w_D t + \varphi), \tag{3}$$

where

$$A = a\omega/\omega_D \text{ and Sin } \varphi = \omega_D/\omega\, .$$

The form (3) for the underdamped solution will be more convenient for our computations. It follows from (3) that

$$x'(t)^2 = A^2\, e^{-2\beta t}\, (\omega^2\, \text{Cos}^2(\omega_D t + \varphi) \tag{4}$$
$$-\, \beta^2\, \text{Cos}2(\omega_D t + \varphi) - \omega_D\beta\, \text{Sin}2(\omega_D t + \varphi))$$

Here we used $\omega^2 - \beta^2 = \omega_D{}^2$ together with the double angle formulas for sine and cosine. Now substituting (3) and (4) into (1) leads to

$$E(t) = 1/2A^2\, e^{-2\beta t}\, (m\omega^2\, \text{Cos}^2(\omega_D t + \varphi) + K\, \text{Sin}^2(\omega_D t + \varphi))$$
$$-1/2A^2\, e^{-2\beta t}\, (\beta^2\, \text{Cos}2(\omega_D t + \varphi) + \omega_D\beta\, \text{Sin}2(\omega_D t + \varphi)).$$

Once again using that $m\omega^2 = K$, this reduces to

$$E(t) = 1/2A^2\, e^{-2\beta t} -$$
$$-1/2A^2\, e^{-2\beta t}\, (\beta^2\, \text{Cos}2(\omega_D t + \varphi) + \omega_D\beta\, \text{Sin}2(\omega_D t + \varphi)).$$

This is the expression for the energy in the spring-mass system in the underdamped case. Note that the second and third terms in the expression for the energy are oscillatory and represent small fluctuations in the energy during each cycle. Then the contribution of these terms to the *average* energy over many cycles is small, and we can approximate $<E(t)>$, the average energy, by

$$<E(t)> \sim 1/2\, KA^2\, e^{-2\beta t} = 1/2Ka^2\, e^{-2\beta t}\, (\omega/\omega_D) \tag{5}$$

It is evident from (5) that the average energy in the system decreases by the factor 1/e in a time period of length $1/(2\beta)$; i.e., the system loses about two thirds of its energy in this period of time. In particular, the total energy is not constant, it is decreasing with time.

These computations demonstrate that damping is an energy dissipating mechanism for the spring-mass system. When no damping is present the energy in the spring-mass system is constant, but when there is damping, energy is dissipated at a rate that increases as the damping increases.

PROBLEM 5.4

A circuit contains a capacitor, an inductor and a variable resistance in series (Figure 5.5.). Such a circuit is called an L-C-R circuit. With the switch in position A, the battery places a charge of Q_O on the capacitor. When the switch is moved to position B, the capacitor discharges into the circuit. Find the resulting current I(t). Find the value of the resitance R in terms of the capacitance C and inductance L such that oscillation in I(t) is just suppressed.

SOLUTION 5.4

When the switch is moved to position B, a charge of Q_O is discharged into a circuit that is initially dead. We describe this situation with the initial conditions

$$Q(0) = Q_O \text{ and } Q'(0) = I(0) = 0. \tag{1}$$

Then Q=Q(t), the charge in the circuit, varies with time $t > 0$ according to the equation

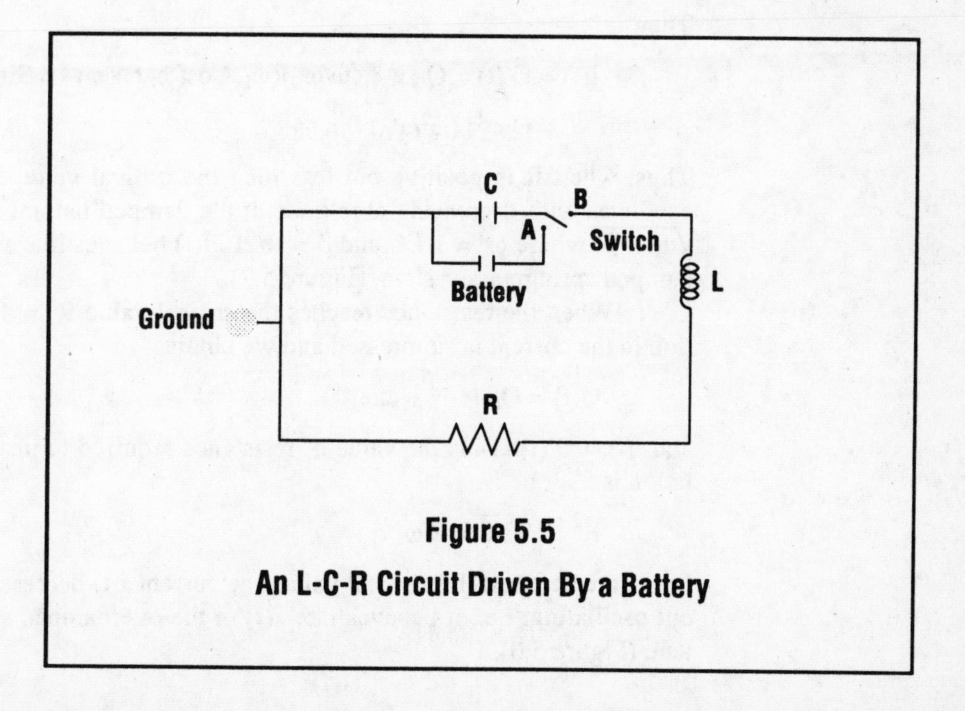

Figure 5.5

An L-C-R Circuit Driven By a Battery

$$LQ''(t) + RQ'(t) + 1/CQ(t) = 0 . \qquad (2)$$

But (1), (2) is the same initial value problem that arose in Problem 5.1 for the spring-mass system with the nominal differences that

- L takes the place of the mass m.
- R takes the place of the damping constant C.
- 1/C takes the place of the spring stiffness K.

Then the solution of (1), (2) can be obtained by using the solutions constructed in Problem 5.1 and replacing m, C, and K by L, R, and 1/C, respectively.

(a) For $R = 0$ (no resistance) :

$$Q(t) = Q_O \operatorname{Cos} \omega t, \qquad \omega = 1/LC,$$
$$I(t) = Q'(t) = -\omega Q_O \operatorname{Sin} \omega t.$$

Note that the natural frequency of the electical circuit is equal to $w=1/\sqrt{LC}$. Thus, when $R=0$, the current oscillates with constant amplitude at the natural frequency ω.

(b) For $0 < R^2 < 4/LC$, the underdamped solution from Problem 5.1 translates to

$$Q(t) = Q_O e^{-\beta t} (\operatorname{Cos} \omega_D t + \frac{\beta}{\omega_D} \operatorname{Sin} \omega_D t)$$
$$= Q_O e^{-\beta t} (\omega/\omega_D) \operatorname{Sin}(\omega_D t + \varphi), \qquad \operatorname{Sin} \varphi = \omega_D/\omega.$$

Then

$$I(t) = Q'(t) = Q_O e^{-\beta t} (\omega/\omega_D)(\omega_D \operatorname{Cos}(\omega_D t + \varphi) - \beta \operatorname{Sin}(\omega_D t + \varphi))$$
$$= Q e^{-\beta t} (\omega^2/\omega_D) \operatorname{Sin} \omega_D t .$$

Thus, when R is positive but less than the critical value $2/\omega$, the current oscillates with decreasing amplitude at the damped natural frequency $w_D = \sqrt{\omega^2 - \beta^2}$ where $\omega^2 = 1/LC$ and $\beta = R/2L$. I(t) behaves like x(t) in the underdamped springmass system, (Figure 5.2).

(c) When the resistance reaches the critical value $R^2 = 4/LC$ the oscillation in the current is suppressed and we obtain

$$Q(t) = Q_O (e^{-\omega t} + \omega t e^{-\omega t}) ,$$

and $I(t) = Q'(t)$. Thus the value of resistance required to just suppress oscillation is

$$R = 2/\sqrt{LC} = 2/\omega.$$

For R greater than this critical value, the current I(t) decreases to zero without oscillating: i.e., it behaves like x(t) in the overdamped spring mass system, (Figure 5.3).

PROBLEM 5.5

A spring-mass system is given an initial displacement equal to a units and is then released from rest. It is observed that the mass completes one full cycle in a time of .35 seconds but, because of damping, returns to a position measured at only .66a units of amplitude. Determine the damping constant C for this system.

SOLUTION 5.5

According to Problem 5.1(b), the amplitude of the oscillating mass is given by

$$x(t) = ae^{-\beta t}\,\mathrm{Sin}(\omega_D t + \varphi). \tag{1}$$

The mass begins in a position of maximum deflection at t=0 and completes one cycle in T=.35 seconds; i.e., $\mathrm{Sin}\varphi = \mathrm{Sin}(\omega_D T + \varphi) = 1$. Then

$$x(T)/x(0) = e^{-bT},$$

and since it is given that $x(T)/x(0) = .66$, we compute

$$\beta = C/2m = -\ln(.66)/.35 = 1.187\ \mathrm{sec}^{-1}.$$

Note that the motion described by (1) is not periodic since x(t) does not return to its initial value after the time period, $T=2\pi/\omega_D$. Instead, x(t) assumes the value $x(T) = x(0)e^{-bT}$, a value that is less than x(0). Nevertheless, we refer to T, the time between successive maximum values for x(t), as the length of a *cycle* or sometimes as a *quasiperiod.*

PROBLEM 5.6

A grandfather clock with a pendulum 100cm in length is losing 5 minutes per hour. How should the length of the pendulum be altered in order for the clock to keep correct time? Assume that the damping can be neglected and the amplitude of the pendulum is sufficiently small that the assumption $\mathrm{Sin}\vartheta(t) = \vartheta(t)$ is valid.

SOLUTION 5.6

If we suppose that the pendulum is given an initial angular displacement ϑ_O and is released from rest, then neglecting friction and assuming small amplitude, the angular displacement as a function of time is given by

$$\vartheta(t) = \vartheta_O\,\mathrm{Cos}\omega t,$$

where $w=\sqrt{g/L}$ denotes the natural frequency of the pendulum; i.e., the period T of the pendulum equals $2\pi/\omega$. If T_O denotes the period the pendulum must have in order to keep correct time, then T_O is related to L_O the correct length by $T_O = 2\pi/\omega_O = 2\pi\sqrt{(L_O/g)}$. But it is given that $T/T_O = 65/60$ so it follows that

$$L_O = (60/65)^2\, L = .85L = 85\text{cm}.$$

Thus, the pendulum should be shortened to a length of 85 cm in order that the period of the pendulum be correspondingly decreased from T to T_O.

Forced Response of Physical Systems

PROBLEM 5.7

A damped spring-mass system is driven by a period forcing term $F(t) = F_O \operatorname{Cos} \Omega t$. If the mass is set in motion with initial displacement equal to x_O and initial velocity of v_O, then find the subsequent motion of the mass. Identify the transient and steady-state components of the solution and estimate the time T after which the amplitude of the transient is less than 1 percent of the amplitude of the steady component.

SOLUTION 5.7

The displacement x(t) of the mass from the equilibrium position satisfies the equation

$$x''(t) + 2\gamma\, x'(t) + \omega^2\, x(t) = F_O/m \operatorname{Cos}\Omega t \qquad (1)$$

and the initial conditions

$$x(0) = x_O, \qquad x'(0) = v_O. \qquad (2)$$

Here γ =C/m and ω^2 =K/m, and we are supposing Ω does not equal ω. Referring to Problem 4.10, a particular solution for (1) is given by,

$$x_p(t) = A \operatorname{Sin} \Omega t + B \operatorname{Cos} \Omega t\,, \qquad (3)$$

where

$$A = \frac{CF_O\Omega}{m^2}\; \frac{1}{(\omega^2 - \Omega^2)^2 + (\gamma\,\Omega)^2}$$

and

$$B = \frac{F_O}{m}\; \frac{\omega^2 - \Omega^2}{(\omega^2 - \Omega^2)^2 + (\gamma\,\Omega)^2}$$

The general solution of (1) is then a function of the form

$$x(t)=e^{-\gamma t}\,(C_1 \operatorname{Sin}\omega_D t + C_2 \operatorname{Cos}\omega_D t) + x_p(t)\,, \qquad (4)$$

for $\omega_D{}^2 = \omega^2 - \gamma^2$. It follows from (4) that $x(0) = C_2 + B$ and $x'(0) = \omega_D C_1 - \gamma C_2 + A\Omega$; using this together with the initial conditions, (3) leads to

$$x(t)=e^{-\gamma t}\,[(v_O - A\Omega)\operatorname{Sin}\omega_D t+(x_O - B)(\gamma \operatorname{Sin}\omega_D t+\omega_D \operatorname{Cos}\omega_D t)]/\omega_D \qquad (5)$$

$$+ A\operatorname{Sin}\Omega t + B\operatorname{Cos}\Omega t$$

Note that x(t) given by (5) consists of the sum of the particular solution $x_P(t)$ and a solution of the homogeneous equation

$$x_H(t)=e^{-\gamma t}\,[(v_O-A\Omega)\mathrm{Sin}\,\omega_D t+(x_O-B)(\gamma\,\mathrm{Sin}\,\omega_D t+\omega_D\,\mathrm{Cos}\,\omega_D t)]/\omega_D. \quad (6)$$

The particular solution represents the response of the spring mass system to the forcing function $F(t)=F_O\,\mathrm{Cos}\,\Omega t$, while the homogeneous solution is added to the particular solution to ensure that x(t) satisfies the initial conditions (3). Hence the homogeneous solution can be viewed as the response of the system to the initial conditions. Note also that $x_H(t)$ tends to zero with increasing t so that the influence of the initial conditions diminishes with time. For this reason we refer to $x_H(t)$ as the *transient* component of the solution. Since $x_H(t)$ decreases to zero as t increases, it follows that x(t) tends to $x_P(t)$ with increasing t. Thus we refer to $x_P(t)$ as the *steady-state* component of the solution. In figure 5.6 we plot x(t) versus t and, in the same figure, the transient solution $x_H(t)$. Note that as $x_H(t)$ decreases toward zero, the solution x(t) becomes more nearly purely sinusoidal; i.e., x(t) tends to $x_P(t)$.

In order to determine the time T at which the amplitude of the transient component of the solution has diminished to just one percent of the amplitude of the steady-state component, we note first that the steady-state amplitude is,

$$AMPL_{SS} = \sqrt{A^2 + B^2}\,.$$

Next, if we write

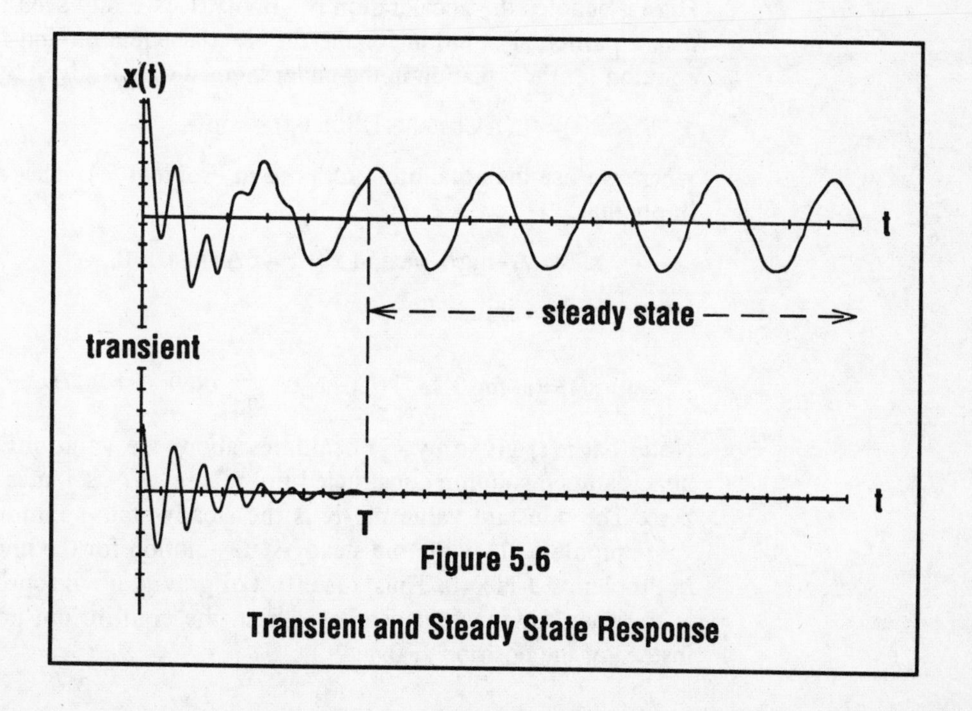

Figure 5.6

Transient and Steady State Response

$$\gamma \text{Sin}\omega_D t + \omega_D \text{Cos}\omega_D t = \omega\,(\text{Cos}\varphi\ \text{Sin}\omega_D t + \text{Sin}\varphi\ \text{Cos}\omega_D t)$$
$$= \omega \text{Sin}(\omega_D t + \varphi),$$

where $\text{Tan}\varphi=\omega_D/\gamma$, then it follows from (6) that the amplitude of the transient D portion of the solution is,

$$\text{AMPL}_{trans} = e^{-\gamma t}\sqrt{(v_O-A\Omega)^2+\omega^2(x_O-B)^2}\,/\omega_D\,.$$

Then the amplitude of the transient component is less than or equal to the amplitude of the steady-state component of the solution for all t greater than or equal to T if T satisfies,

$$e^{-\gamma T} = .01\omega_D\sqrt{A^2+B^2}\,/\sqrt{(v_O-A\Omega)^2+\omega^2(x_O-B)^2}\,.$$

This equation may be solved for T by taking the logarithm of both sides.

PROBLEM 5.8

Determine the effect of the force of gravity on the underdamped motion of the spring-mass system of Problem 5.1.

SOLUTION 5.8

If the mass is given an initial displacement and is released from rest under the influence of the force of gravity, then the displacement x(t) solves the initial value problem

$$mx''(t)+Cx'(t)+Kx(t)=mg, \qquad x(0)=a,\ x'(0)=0. \tag{1}$$

Here g denotes the acceleration of gravity. It is easily seen that $x_P(t)=mg/K$ p is a particular solution for the differential equation and hence a general solution for the equation in the underdamped case is given by

$$x(t)={}^{-\beta t}(A\text{Cos}\omega_D t+B\text{Sin}\omega_D t) + mg/K,$$

where we use the notation introduced in Problem 5.1. The initial conditions imply that

$$x(0)= A+mg/K=a \text{ and } x'(0)=-\beta A+\omega_D tB=0,$$

leading to the solution

$$x(t)=(a-mg/K)e^{-\beta t}\,(\text{Cos}\omega_D t+\frac{\beta}{\omega_D}\text{Sin}\omega_D t) + mg/K\,. \tag{2}$$

Note that x(t) given by (2) oscillates about the value mg/K whereas the underdamped solution constructed in Problem 5.1 oscillates about the value zero. The constant value mg/K is the steady-state solution for the initial value problem (1) while the steady-state solution for the underdamped case in Problem 5.1 is x=0. Thus, the effect of gravity on the spring mass system is to cause the mass to oscillate about the equilibrium position x=mg/K instead of the position x=0.

PROBLEM 5.9

An undamped spring-mass system driven by a periodic forcing function is modeled by the initial value problem

$$x''(t) + \omega^2 x(t) = (F/m)\text{Cos}\Omega t$$
$$x(0) = x'(0) = 0.$$

(a) Show that when Ω is not equal to ω, then $x(t)$ can be written as

$$x(t) = \frac{2\ F/m}{\omega^2 - \Omega^2}\text{Sin}[(\Omega-\omega)t/2]\ \text{Sin}[(\Omega+\omega)t/2] \tag{1}$$

and, by plotting $x(t)$ versus t, illustrate the phenomenon of *beats.*

(b) Take the limit as W approaches w and show that one obtains the *resonant* solution

$$x(t) = \frac{F}{2m\omega}\ t\ \text{Sin}\omega t\ . \tag{2}$$

SOLUTION 5.9

(a) When Ω is not equal to ω, the general solution of the initial value problem is given by

$$x(t) = A\ \text{Cos}\omega t + B\ \text{Sin}\omega t + \frac{F/m}{\omega^2-\Omega^2}\text{Cos}\Omega t\ .$$

Since, $x(0)=A+ F/m(\omega^2-\Omega^2)$ and $x'(0)=B\omega$, the initial conditions lead to the following unique solution for the initial value problem:

$$x(t) = \frac{F}{m(\omega^2-\Omega^2)}(\text{Cos}\Omega t-\text{Cos}\omega t)\ . \tag{3}$$

If we define φ and σ by

$$\varphi=(\Omega+\omega)/2 \text{ and } \sigma=(\Omega-\omega)/2,$$

then

$$\Omega= \varphi+\sigma \text{ and } \omega=\varphi-\sigma\ .$$

Then the identities for the cosine of a sum and difference of two angles lead to

$$\text{Cos}\Omega t = \text{Cos}\varphi t\ \text{Cos}\sigma t - \text{Sin}\varphi\text{Sin}\sigma t$$

$$\text{Cos}\omega t = \text{Cos}\varphi t\ \text{Cos}\sigma t + \text{Sin}\varphi t\ \text{Sin}\sigma t$$

and

$$x(t) = -2\,F/(m(\omega^2-\Omega^2))\ \text{Sin}\varphi t\,\text{Sin}\sigma t\,.$$

That is,

$$x(t) = \frac{2\,F/m}{\Omega^2-\omega^2}\ \text{Sin}((\Omega-\omega)t/2)\ \text{Sin}((\Omega+\omega)t/2)$$

Here we see that x(t) can be viewed as

$$x(t) = A(t)\text{Sin}\varphi t;$$

i.e., x(t) is oscillating at the "fast frequency" $\varphi=(\Omega + \omega)/2$ with variable amplitude A(t), given by

$$A(t)=2F/[m(\Omega^2-\omega^2)]\ \text{Sin}\sigma t$$

which oscillates at the "slow frequency" $\sigma=(\Omega-\omega)/2$. We have plotted x(t) versus t in figure 5.7(a), where it can be seen that maximum values of amplitude are separated by intervals of length $T=2\pi/\sigma$. In acoustical applications, periodical variations in amplitude are referred to as *beats.* There the periodic variations are audible, with the loudest sound occuring at the peak amplitudes.

(b) As the forcing frequency Ω approaches ω, the natural frequency of the system, we approach a condition of *resonance.* Letting Ω tend to ω in (3) leads to an indeterminate form which may be evaluated by means of L'Hospital's rule to obtain

$$\lim_{\Omega\to\omega} x(t) = \frac{F}{m}\ \frac{-t\,\text{Sin}\Omega t}{-2\Omega}$$

That is, when Ω equals ω, x(t) given by (3) reduces to,

$$x(t) = \frac{F}{2m\omega}\,t\ \text{Sin}\omega t\,. \tag{4}$$

This is in agreement with the resonant case solution constructed in Problem 4.12 by means of undetermined coefficients. The resonant solution (4) is plotted in Figure 5.7(b), where it is evident that the amplitude of an undamped system in a state of resonance increases without bound.

In the next problem we demonstrate the effect of damping on resonance.

PROBLEM 5.10

A damped spring-mass system is driven by a periodic forcing function. The displacement x(t) satisfies

$$mx''(t) + Cx'(t) + Kx(t) = F\text{Cos}\ \Omega t, \tag{1}$$

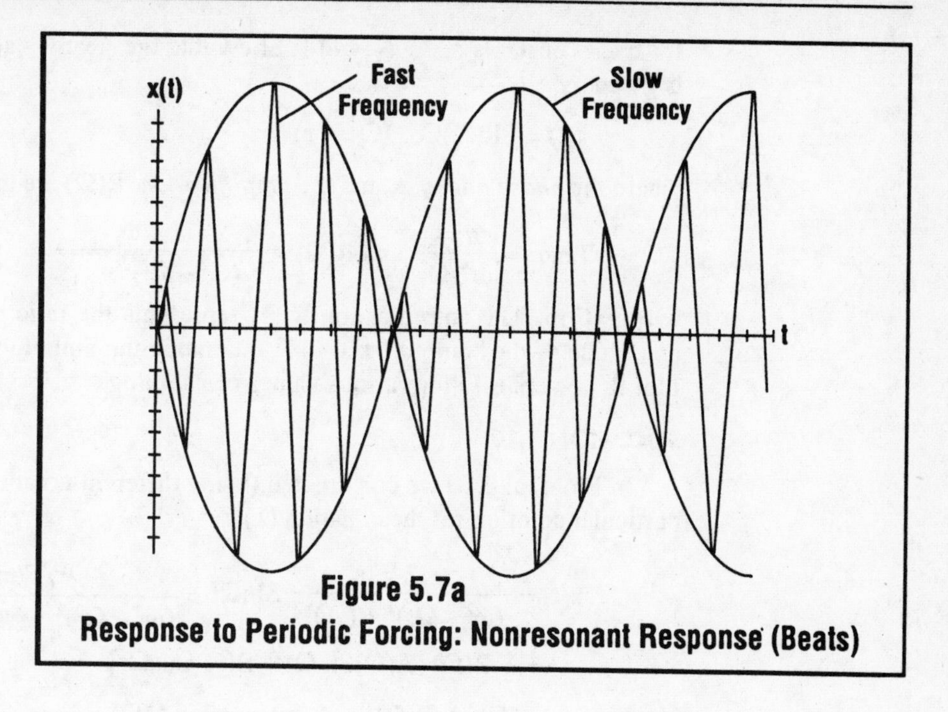

Figure 5.7a
Response to Periodic Forcing: Nonresonant Response (Beats)

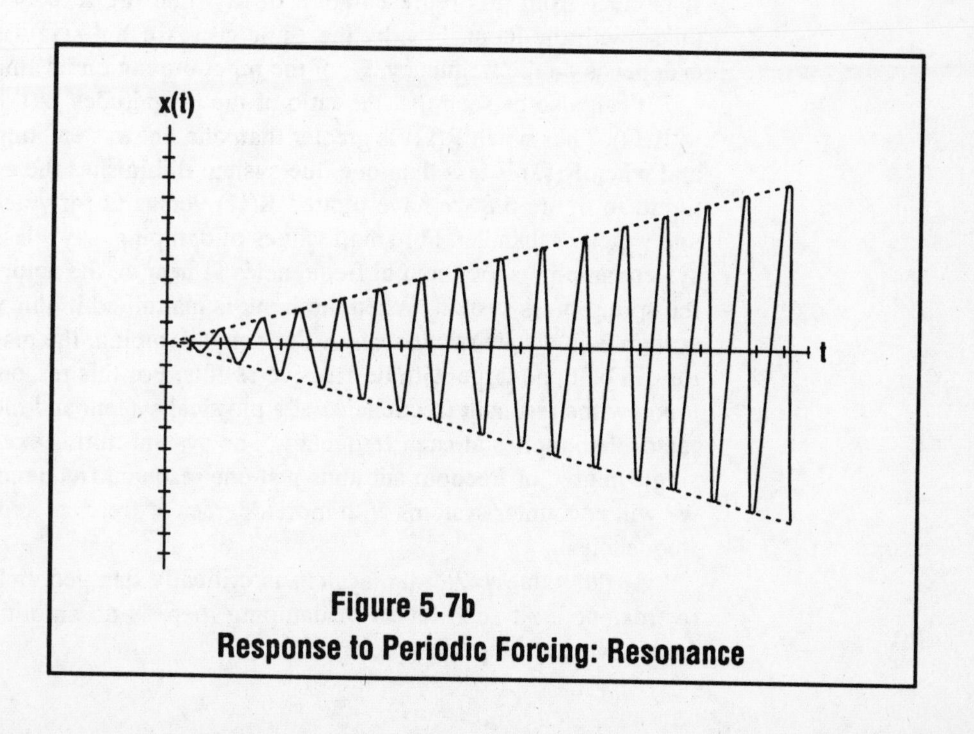

Figure 5.7b
Response to Periodic Forcing: Resonance

for fixed constants m, C, K, and F. Show that the steady-state solution of (1) is given by

$$x(t) = FR(\Omega)\mathrm{Cos}\,(\Omega t - \varphi) \tag{2}$$

where the *phase angle,* φ, and the *amplification,* R(Ω) are given by

$$\mathrm{Tan}\varphi = \frac{\gamma\,\Omega}{\omega^2 - \Omega^2} \text{ and } R(\Omega) = \frac{1/m}{\sqrt{(\omega^2 - \Omega^2)^2 + (\gamma\Omega)^2}}$$

with $\gamma = C/m$. The amplification R(Ω) represents the ratio of the maximum amplitude of the "output", x(t) to F, the maximum amplitude of the "input". Plot R(Ω) versus Ω for various values of damping.

SOLUTION 5.10

In Problem 4.10 we constructed (using different notation) the following particular solution for the equation (1):

$$x(t) = \frac{\gamma\,\Omega\, F/m}{(\omega^2 - \Omega^2)^2 + (\gamma\Omega)^2}\mathrm{Sin}\Omega t + \frac{(\omega^2 - \Omega^2)\, F/m}{(\omega^2 - \Omega^2)^2 + (\gamma\Omega)^2}\mathrm{Cos}\Omega t$$
$$= F\,R(\Omega)\,[A(\Omega)\mathrm{Sin}\Omega t + B(\Omega)\mathrm{Cos}\Omega t].$$

Noting that $A(\Omega)^2 + B(\Omega)^2 = 1$, we can let $A(\Omega) = \mathrm{Sin}\varphi$ and $B(\Omega) = \mathrm{Cos}\varphi$ so that the expression for x(t) reduces to

$$x(t) = FR(\Omega)\mathrm{Cos}(\Omega t - \varphi).$$

It is clear from this representation of x(t) that the response x(t) is out of phase with the input, F(t), by the angle $\varphi = \mathrm{ArcTan}[A(\Omega)/B(\Omega)]$. Note that φ depends on the frequency, Ω, of the input but not on its amplitude, F.

It can also be seen that the ratio of the magnitudes $\|x(t)\|/\|F(t)\|$ is equal to R(Ω). Thus when R(Ω) is greater than one, the system amplifies the input and when R(Ω) is less than one, the system diminishes the amplitude of the input. In figure 5.8 we have plotted R(Ω) versus Ω for $\gamma = \omega/2$, ω, $\omega\sqrt{2}$, 2ω and γ greater than 2ω. For small values of damping, say γ less than $\omega\sqrt{2}$, the system amplifies the input at frequencies Ω near ω, the natural frequency of the spring-mass system. When the input is magnified in this way we say the system is in a state of *resonance*. For small damping, the magnification factor can be quite large with destructive results. For this reason it is important to know the resonant frequencies of a physical system and the magnification factor that applies at each frequency. The system in this example has just a single degree of freedom and thus just one resonant frequency. In Chapter 7 we will encounter systems with more degrees of freedom and more resonant frequencies.

At the value $\gamma = 2\omega$, the system is critically damped and for $\gamma > 2\omega$ it is overdamped. At such levels of damping there is no amplification of input for any value of input frequency, Ω.

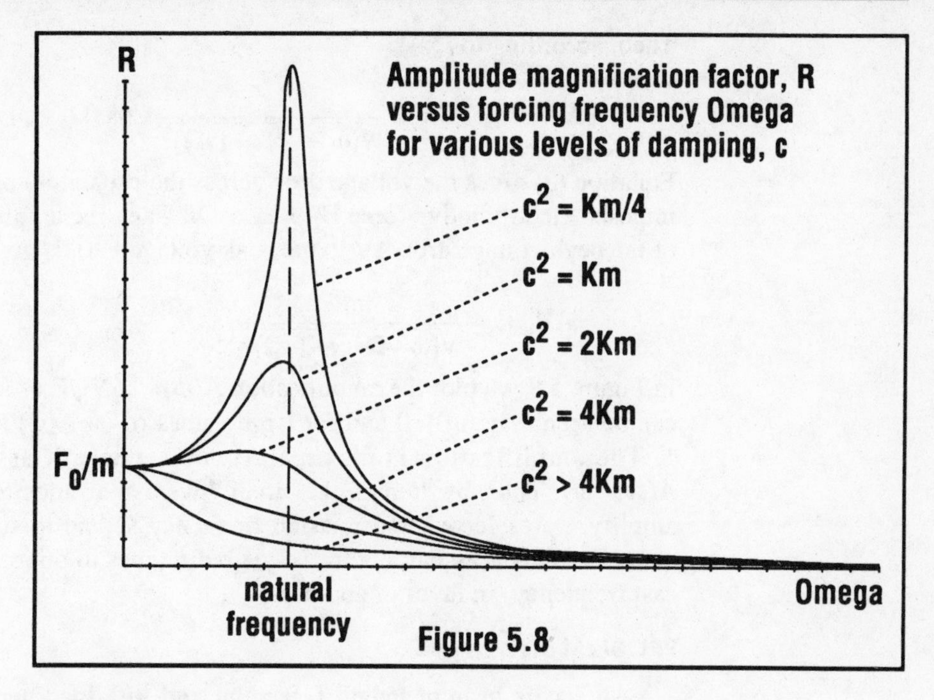

Figure 5.8

PROBLEM 5.11

An electric circuit contains in series an inductance, a resistance, and a variable capacitance. The circuit is driven by a periodic electromotive force, $E(t)=E\text{Cos}\Omega t$. Compute ΔV, the voltage drop across the variable capacitor as a function of t. Show that as the capacitance varies, the amplification, $\|\Delta V\|/E$, varies and plot the amplification as a function of $\omega=1/\sqrt{LC}$.

SOLUTION 5.11

The charge Q(t) in the circuit solves the equation

$$LQ''(t) + RQ'(t) + 1/C\, Q(t) = E(t). \tag{1}$$

Note that except for minor changes in notation, equation (1) is the equation that appears in Problem 4.10. Then, using the particular solution constructed in Problem 4.10, we write

$$Q(t) = \frac{\gamma\Omega\, E/L}{(\omega^2-\Omega^2)^2 + (\gamma\Omega)^2}\text{Sin}\Omega t + \frac{(\omega^2-\Omega^2)\, E/L}{(\omega^2-\Omega^2)^2 + (\gamma\Omega)^2}\text{Cos}\Omega t \tag{2}$$

$$= E/L \frac{1}{\sqrt{(\omega^2-\Omega^2)^2 + (\gamma\Omega)^2}}\text{Cos}(\Omega t - \varphi)$$

Here we have let

$$\gamma = R/L \text{ and Tan } \varphi = \frac{\gamma\Omega}{\sqrt{(\omega^2-\Omega^2)^2 + (\gamma\Omega)^2}}\,.$$

Then, according to (5.4),

$$\Delta V = Q(t)/C = E \frac{\omega^2}{\sqrt{(\omega^2-\Omega^2)^2 + (\gamma\Omega)^2}} \operatorname{Cos}(\Omega t - \varphi) . \tag{3}$$

Equation (3) gives the voltage drop across the capacitor corresponding to an imposed electromotive force $E(t)=E\operatorname{Cos}\Omega t$. Then the amplification, the ratio of the peak voltage drop ΔV to the peak voltage $E(t)$, is given by

$$\|\Delta V\|/E = \frac{\omega^2}{\sqrt{(\omega^2-\Omega^2)^2 + (\gamma\Omega)^2}} . \tag{4}$$

In Figure 5.9 we plot the amplification $A(\omega)= \|\Delta V\|/E$ as a function of ω. It can be seen that $A(0)=0$ and for large values of ω, $A(\omega)$ tends to the value 1. The amplification is maximal at $\omega=\Omega$ where it assumes the value $A(\Omega)=\omega/\gamma$. Thus, by "tuning the circuit" we can arrange to have the circuit amplify a preselected transmission frequency Ω and to suppress other frequencies. This is essentially the way a radio tunes to one station (i.e., broadcast frequency) in favor of another.

PROBLEM 5.12

An elastic beamof length L is subjected to a load per unit length $F(x)$ that increases linearly with x along the beam; i.e., $F(x) = \lambda x$.

(a) Find $y(x)$, the deflection of the centerline of the beam if the end $x=0$ is rigidly clamped while the end $x=L$ is free and unsupported.

(b) Find $y(x)$ in the case that both ends of the beam are simply supported.

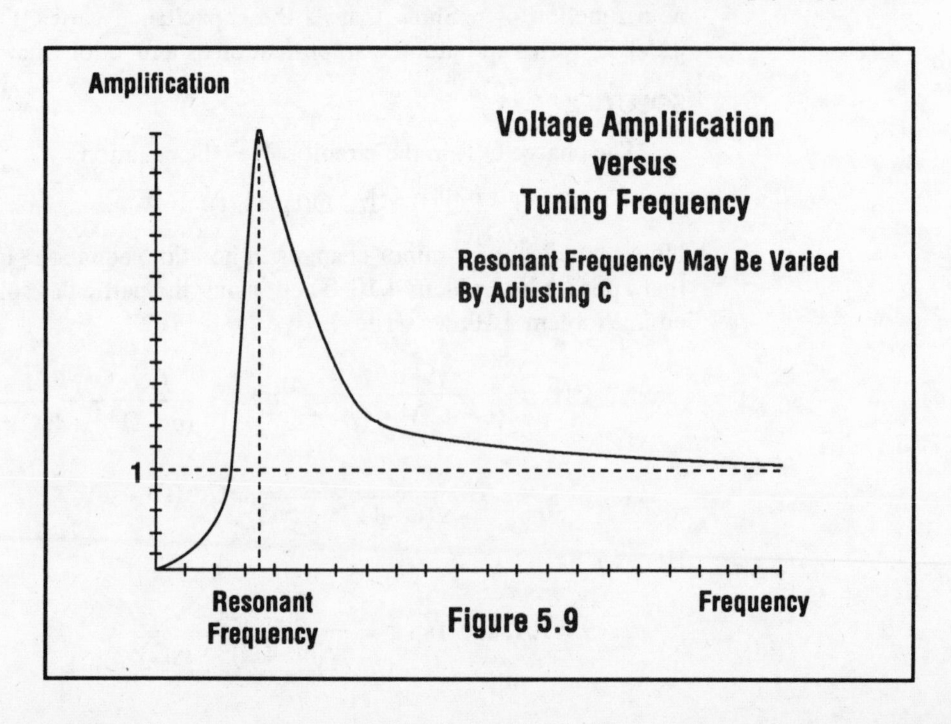

Figure 5.9

SOLUTION 5.12

(a) For $F(x) = \lambda x$, $y(x)$, the deflection of the centerline, satisfies

$$y^{(4)}(x) = \lambda x/EI. \quad (1)$$

In addition, we express the fact that the beam is rigidly clamped at the end $x=0$ by

$$y(0)=0 \text{ and } y'(0)=0\,. \quad (2)$$

From (5.9) and (5.10) it follows that if the end $x=L$ is free (i.e., no shear and no moment), then

$$y''(L)=0 \text{ and } y^{(3)}(L)=0. \quad (3)$$

The auxiliary equation associated with (1) has zero as a root of multiplicity four. It follows then from Theorems 4.4 and 4.5 that (1) has a homogeneous solution of the form

$$y_H(x)= A + Bx + Cx^2 + Dx^3$$

and a particular solution of the form

$$y_P(x)=Rx^4\,.$$

The method of undetermined coefficients leads to the result $R = \lambda/24EI$. Then the general solution of (1) is given by

$$y(x)= A + Bx + Cx^2 + Dx^3 + \lambda x^4/24EI.$$

The boundary conditions (2) imply

$$y(0) = A = 0 \text{ and } y'(0) = B = 0.$$

Similarly, the boundary conditions (3) imply that

$$y''(L)= 2C + 6DL + \lambda L^2/2EI = 0$$

and

$$y^{(3)}(L)= 6D + \lambda L/EI = 0,$$

which leads to the solution

$$y(x) = \lambda x^2(x^2-4xL+6L^2)/24EI. \quad (4)$$

Using (4), not only can we determine the deflection of the centerline of the beam, we can also compute the distribution of shear and moment in the beam as a function of x.

(b) If the beam is simply supported at each end, then the ends of the beam are prevented from deflecting up or down but are free to rotate. The boundary conditions describing this state of affairs are

$$y(0) = y(L) = 0 \text{ (zero deflection at each end)}$$

and

$$y''(0) = y''(L) = 0 \text{ (zero moment at each end)}$$

As before, equation (1) has a general solution of the form

$$y(x) = A + Bx + Cx^2 + Dx^3 + \lambda x^4/24EI.$$

The boundary conditions imply that

$$y(0)=A=0 \text{ and } y''(0)=2C=0$$

$$y(L)=BL+DL^3+\lambda L^4/24EI=0$$

$$y''(L)=6DL+\lambda L^2/2EI=0,$$

and this leads easily to the result

$$y(x)=\lambda x(L^3-2Lx^2+x^3)/24EI. \tag{5}$$

The expression (5) not only defines the deflection of the centerline of the beam under these conditions of loading and support but permits computation of the distributions of moment and shear in the beam.

In this chapter we saw that there are several physical systems whose behavior is described by the second order linear equation, $Ay''(t) + By'(t) + Cy(t) = 0$. With the appropriate interpretation of the coefficients A,B,C and the unknown function y(t), these include:

- longitudinal spring $mx''(t)+Cx'(t)+Kx(t)=F(t)$
- torsional spring $I\vartheta''(t)+C\vartheta'(t)+K\vartheta(t)=T(t)$
- pendulum (small amplitude motion) $mL\vartheta''(t)+C\vartheta'(t)+mg\vartheta(t)=0$
- electrical circuit $LI''(t)+RI'(t)+1/CI(t)=E'('t)$

A solution of an initial value problem for the homogeneous differential equation is called a free response. *The nature of the free response is determined by the coefficient of the first derivative term. In the case of the spring-mass system, the coefficient of the first derivative term is the damping. If the damping is equal to zero then the solution oscillates with constant amplitude at the* natural frequency *of the spring. If the damping is greater than zero, but less than the critical value, then the solution oscillates with steadily decreasing amplitude at the* damped natural frequency *which is less than the undamped natural frequency. The critical value of the damping is the smallest value of the coefficient of the first derivative term for which no oscillation occurs. For damping equal to or greater than the critical value,* the solution does not oscillate but instead tends to zero with increasing time. The damping is an energy-dissipating mechanism in the physical system. In a system with zero damping, the energy is constant but in a system with damping, the energy decreases steadily with time.

A solution of an initial value problem for the inhomogeneous differential equation is called a forced response. The general solution to the inhomogeneous equation consists of the sum of a particular solution and a general homogeneous solution. Then the solution to the initial value problem is obtained when the arbitary constants in the homogeneous solution are chosen such that the initial conditions are satisfied. If the coefficient of the first derivative term in the equation is positive then the homogeneous solution tends to zero with increasing t; i.e., this is the transient solution. The particular solution is the steady-state component of the solution.

If the forcing function is periodic, then the steady state solution is also periodic with the same frequency. If the frequency of the periodic forcing function is equal to the natural frequency of the physical system then resonance occurs. If there is no damping, the response is periodic with a steadily increasing amplitude. If the system is damped, then the output has the same frequency as the input, but the phase is shifted and the amplitude may be magnified. In an LCR circuit, we can install a variable capacitance so as to amplify a preselected input frequency. If the input frequency is near but not equal to the resonant frequency, beats may be observed.

The transverse vibration of an elastic beam is one example of a physical system that is modeled by a differential equation of order higher than two.

6

Laplace Transform Methods

In Chapter four we saw that solving a constant coefficient, linear differential equation of order N reduces to the algebraic problem of finding the roots of an Nth degree polynomial. The Laplace transform provides a means for systematically transforming an initial value problem for a linear, constant coefficient differential equation into an associated algebraic problem. The algebraic problem is presumably more easily solved, and the solution of the algebraic problem then leads to the solution of the initial value problem by inverting the Laplace transform. In addition, the Laplace transform provides a simple way of studying the relation of system output to system input and of examining system stability.

In the first two sections, we develop the rudimentary Laplace transform techniques required to solve typical initial value problems. In the next section, we introduce the notions of the step function and impulse function and compute their transforms. All of this is put to use in the last section, where we present several examples designed to illustrate the advantages and disadvantages of the Laplace transform.

THE LAPLACE TRANSFORM

We shall consider functions f(t),g(t),... defined for $t \geq 0$, and define the Laplace transform of f(t) as

$$F(s) = \int_0^\infty f(t)\, e^{-st}\, dt. \tag{6.1}$$

Occasionally we shall use the alternative notation £{f} to indicate the Laplace transform of f(t). The functions f(t) and F(s) are referred to as a *Laplace transform pair.*

Example 6.1 Laplace transform formulas

6.1(A)

Consider the function f(t) = 1 for $t \geq 0$. Then, according to (6.1),

$$£\{1\} = \int_0^\infty 1\, e^{-st} dt = \left.\frac{e^{-st}}{-s}\right|_0^\infty = \frac{1}{s}\,.$$

The improper integral converges for $s > 0$ and we conclude that F(s)= 1/s for $s > 0$ in the case of the function f(t) = 1.

6.1(B)

For k a real number, consider $f(t) = e^{kt}$. Then

$$£\{e^{kt}\} = \int_0^\infty e^{-(s-k)t}\, dt = \left.\frac{e^{-(s-k)t}}{-(s-k)}\right|_0^\infty = \frac{1}{s-k}\,.$$

Here, the improper integral converges for $s > k$ and hence $£\{e^{kt}\}$ is equal to $1/(s-k)$ for $s>k$.

Functions of Exponential Type

In order for the improper integral (6.1) to converge, it is sufficient that there exist constants M and b such that,

$$|f(t)| \leq M\, e^{bt} \text{ for } t \geq 0. \tag{6.2}$$

Any function f(t) that satisfies (6.2) is said to be of *exponential type.* In particular, polynomials in t, Sin t, Cos t are all of exponential type. We shall suppose, in addition, that f(t) is *piecewise continuous;* i.e., f(t) is piecewise continuous if on any interval I of finite length in $[0,\infty)$, f(t) is continuous on I, except possibly at a finite number of finite jump discontinuities.

For any f(t) that is of exponential type and piecewise continuous on $[0,\infty)$, there exists a positive number σ such that the improper integral in (6.1) converges for $s>\sigma$. Then F(s) is a well defined function of s on $[\sigma,\infty)$. For most purposes, it is not necessary that we know σ explicitly and we shall not mention σ unless it is necessary to do so.

Additional Laplace transform formulas could now be computed using (6.1) for other choices of f(t). However, it will be easier to generate transform pairs using the so called operational properties of the transform.

PROPERTIES OF THE LAPLACE TRANSFORM

In example 6.1 we found the Laplace transform for two specific functions. In this section we derive a number of properties of the Laplace transforms of all functions f,g that are piecewise continuous and of exponential type.

Linearity and Homegeneity

For arbitrary functions f and g and all constants a and b,

$$\mathcal{L}\{\, af+bg \,\} = a\,\mathcal{L}\{f\} + b\,\mathcal{L}\{g\} \tag{6.3}$$

$$\mathcal{L}\{f(at)\} = (1/a)\,F(s/a) \text{ for a not zero.} \tag{6.4}$$

Laplace Transform of the Derivative

If f and f′ are piecewise continuous and of exponential type, then

$$\mathcal{L}\{f'\} = sF(s) - f(0). \tag{6.5}$$

If f,f′, and f″ are piecewise continuous and of exponential type, then

$$\mathcal{L}\{f''\} = s^2F(s) - sf(0) - f'(0), \text{ etc.} \tag{6.6}$$

Derivative of the Laplace Transform

For arbitrary f of exponential type

$$\begin{aligned} &\mathcal{L}\{\, tf(t) \,\} = -F'(s) \\ &\mathcal{L}\{\, t^2f(t) \,\} = F''(s), \text{ etc.} \end{aligned} \tag{6.7}$$

Note that properties (6.5) and (6.7) are, in a sense, duals of one another. Property (6.5) states that differentiating f(t) corresponds to multiplying F(s) by the transform variable, s. Conversely, multiplying f(t) by the time variable, t, corresponds to differentiation of F(s).

Example 6.2 Further Laplace transform formulas

6.2(A)

For f(t) = 1, we have F(s) = 1/s. Then using property (6.7), we find

$$\mathcal{L}\{t\} = \mathcal{L}\{tf(t)\} = -F'(s) = 1/s^2;$$

i.e., for f(t)=t, we have F(s)= $1/s^2$. Now apply (6.7) again with this choice of f and F:

$$\mathcal{L}\{t^2\} = \mathcal{L}\{tf(t)\} = -F'(s) = 2/s^3.$$

Repeating this process again leads to

$$\mathcal{L}\{t^3\} = \mathcal{L}\{tf(t)\} = -F'(s) = 6/s^4,$$

and, in general, for any integer N,

$$\mathcal{L}\{t^N\} = \mathcal{L}\{tf(t)\} = -F'(s) = N!/s^{N+1}.$$

6.2(B)

For $f(t) = \text{Sin } t$, we have $f'(t) = \text{Cos } t$ and $f''(t) = -f(t)$. Then, using property (6.6), we have

$$£\{-\text{Sin } t\} = -F(s) = s^2F(s) - sf(0) - f'(0).$$

Since $f(0)=0$ and $f'(0)=1$, we can solve for $F(s)$ to obtain

$$£\{\text{Sin } t\} = F(s) = \frac{1}{s^2 + 1}$$

In addition, since $f'(t) = \text{Cos } t$,

$$£\{\text{Cos } t\} = s\,F(s) - f(0) = \frac{s}{s^2 + 1}$$

6.2(C)

Apply property (6.4) to the results just computed to obtain

$$£\{\text{Sin } at\} = \frac{1}{a}\,\frac{1}{(s/a)^2+1} = \frac{a}{s^2 + a^2}$$

and

$$£\{\text{Cos } at\} = \frac{1}{a}\,\frac{s/a}{(s/a)^2+1} = \frac{s}{s^2 + a^2}$$

Now apply the property (6.7) to this result,

$$\begin{aligned} £\{t\text{ Sin } at\} &= -\, d/ds(a/(s^2+a^2)) \\ &= 2as/(s^2+a^2)^2 \end{aligned}$$

and

$$\begin{aligned} £\{t\text{ Cos } at\} &= -\, d/ds(s/(s^2+a^2)) \\ &= (s^2-a^2)/(s^2+a^2)^2. \end{aligned}$$

We consider now additional properties of the Laplace transform.

Shifting the Transform

For arbitrary real constant b,

$$£\{e^{bt}f(t)\} = F(s-b) \qquad (6.8)$$

SHIFTING THE FUNCTION

For arbitrary positive constant b,

$$\pounds\{H(t-b)f(t-b)\} = e^{-bs}\,F(s). \tag{6.9}$$

Here H(t) denotes the *Heaviside step function:*

$$H(t) = \begin{cases} 0 \text{ if } t < 0 \\ 1 \text{ if } t > 0\,. \end{cases}$$

Then H(t–b) is zero for t<b and is equal to 1 for t>b. The function H(t–b)f(t–b) is zero for t<b and for t>b it is the function f(t) "delayed by b units"; i.e., the graph of the delayed function is just the graph of f(t) shifted to the right by b units.

Example 6.3 The Shifting Properties of the Transform

6.3(A)

Using (6.8) and previously derived transform formulas, we obtain the following additional formulas:

$$\pounds\{te^{bt}\} = 1/(s-b)^2$$

$$\pounds\{e^{bt}\text{Sin at}\} = \frac{a}{(s-b)^2+a^2}$$

and

$$\pounds\{e^{bt}\text{Cos at}\} = \frac{s-b}{(s-b)^2+a^2}$$

6.3(B)

Consider the following piecewise constant function,

$$f(t) = \begin{cases} 1 \text{ for } 0 < t < 1 \\ -2 \text{ for } 1 < t < 6 \\ 3 \text{ for } t > 6 \end{cases}$$

Then f(t) can be expressed as the following combination of step functions:

$$f(t) = 1 - 3H(t-1) + 5H(t-6).$$

It follows from (6.9) that f(t) has as its transform

$$F(s) = (1/s)(1 - 3e^{-s} + 5e^{-6s}).$$

This transform formula can also be obtained directly by using the definition.

The Convolution Product

Finally, if F(s) and G(s) denote the Laplace transforms of f(t) and g(t), then we shall need to know what function of t has as its transform the product F(s)G(s); it is not the pointwise product f(t)g(t).

CONVOLUTION

For f(t),g(t) defined and piecewise continuous on $[0,\infty)$ let $f*g(t)$ denote the function

$$f*g(t) = \int_0^\infty f(\tau)\, g(t-\tau)\, d\tau \qquad (6.10)$$

We refer to $f*g$ as the *convolution product* of f and g. Note that since $g(t-\tau)=0$ for $\tau>t$, the interval of integration in (6.10) need only extend from 0 to t.

It is easy to show that the convolution product is *commutative* and *distributive*;i.e.,

$$f*g(t) = g*f(t)$$

$$f*[g+h](t) = f*g(t) + f*h(t).$$

The significance of the convolution product for the Laplace transform lies in the following result.

Products of Transforms

$$\pounds[f*g] = F(s)G(s) \qquad (6.11)$$

Property (6.11) of the Laplace transform will be frequently useful in connection with finding *inverse Laplace transforms;* i.e., for finding h(t) given H(s). Generally the inversion of the Laplace transform amounts to looking it up in a table of transforms. Table 6.1 lists commonly used Laplace transform formulas, including all those derived in the examples. When we wish to invert a transform, we look in the right column of Table 6.1 until we find the tranform whose inverse we seek; the inverse is then listed opposite the transform in the left column of Table 6.1. When H(s)=F(s)G(s), where F(s) and G(s) appear in Table 6.1, then property (6.11) implies that $h(t)=f*g(t)$. This situation occurs frequently in the applications.

Direct methods do exist for computing inverse Laplace transforms without reference to tables. However, these methods are beyond the scope of this text.

For convenient reference, Table 6.2 collects the operational properties of the Laplace transform, properties (6.3) through (6.11).

TABLE 6.1 LAPLACE TRANSFORM FORMULAS

	f(t)	F(s)
1.	1	$1/s$
2.	e^{kt}	$1/(s-k)$
3.	t^N for N=integer>0	$N!/s^{N+1}$
4.	Sin at	$\dfrac{a}{s^2+a^2}$
5.	Cos at	$\dfrac{s}{(s^2+a^2)}$
6.	t Sin at	$\dfrac{2as}{(s^2+a^2)^2}$
7.	t Cos at	$\dfrac{s^2-a^2}{(s^2+a^2)^2}$
8.	$t^N e^{bt}$	$N!/(s-b)^{N+1}$
9.	e^{bt}Sin at	$\dfrac{a}{(s-b)^2+a^2}$
10.	e^{bt}Cos at	$\dfrac{s-b}{(s-b)^2+a^2}$
11.	$H(t-T)$	e^{-sT}/s
12.	$\delta(t-T)$	e^{-sT}

STEP FUNCTIONS AND IMPULSE FUNCTIONS

Heaviside Step Function

We define the Heaviside step function H(t) to be the following piecewise continuous function:

$$H(t) = \begin{cases} 0 \text{ if } t < 0 \\ 1 \text{ if } t > 0 . \end{cases} \tag{6.12}$$

TABLE 6.2 PROPERTIES OF THE LAPLACE TRANSFORM

	f(t)	F(s)
1.	$A\,f(t) + B\,g(t)$	$A\,F(s) + B\,G(s)$
2.	$f(at)$ (a not zero)	$1/a\,F(s/a)$
3.	$f'(t)$ $f''(t)$ . . $f^{(N)}(t)$	$s\,F(s) - f(0)$ $s^2\,F(s) - sf(0) - f'(0)$ $s^N\,F(s) - s^{N-1}f(0) \ldots - f^{(N-1)}(0)$
4.	$t\,f(t)$ $t^2 f(t)$ . . $t^N f(t)$	$-F'(s)$ $F''(s)$ $(-1)^N F^{(N)}(s)$
5.	$e^{bt}\,f(t)$	$F(s{-}b)$
6.	$H(t-T)\,f(t-T)$ $T>0$	$e^{-sT}F(s)$
7.	$f{*}g(t) = \int_0^t f(t-\tau)g(\tau)\tau$	$F(s)\,G(s)$

Then for any $T > 0$, $H(t-T)$ is the function that is zero for $0 < t < T$ and is equal to 1 for $t > T$. Sometimes, $H(t-T)$ is referred to as the *unit step function at* t=T.

We have, $\pounds\{H(t)\} = 1/s$ and, by property (6.9) of the transform, $\pounds\{H(t-T)\} = e^{-sT}/s$.

Unit Impulse Function

We shall also have need of the *unit impulse function* which we denote by $\delta(t)$. We shall define the unit impulse to be the "function" whose Laplace transform is equal to 1; i.e.,

$$\pounds\{\delta(t)\} = 1 \qquad (6.13)$$

Generalized Functions

In fact, there is no function, in the classical sense of the term, whose Laplace transform is equal to 1. However, there is a mathematically consistent theory of *generalized functions* in which $\delta(t)$ is perfectly well defined. We shall be able to make use of the "δ-function" in a formal way without involving ourselves in this theory.

If $\delta(t)$ satisfies (6.13), then property (6.9) of the transform implies that for each $T>0$,

$$\pounds\{\delta(t-T)\} = e^{-sT}, \tag{6.14}$$

and property (6.10) implies that for all piecewise continuous functions $f(t)$

$$\delta * f(t) = \int_0^\infty \delta(t-\tau)f(\tau)d\tau = f(t). \tag{6.15}$$

Finally, property (6.5) implies

$$\delta(t-T) = H'(t-T). \tag{6.16}$$

The results (6.13) through (6.16) are all *formal* in the sense that they cannot be justified within the framework of classical analysis. However, in the setting of generalized functions they can be rigorously justified and we shall feel entitled to make use of all of these results.

Results (6.15) and (6.16) in particular, support the intuitive impression that $\delta(t-T)$ is zero everywhere except at $t=T$ where it becomes infinite. This impression is mathematically incorrect but suggests why we use $\delta(t-T)$ to model the physical notion of an *impulse* acting at $t=T$.

APPLICATION OF THE LAPLACE TRANSFORM

Linear Differential Equations

Consider the following linear initial value problem for the unknown function $x=x(t)$:

$$Ax''(t) + Bx'(t) + Cx(t) = f(t) \text{ for } t>0,\ x(0) = P,\ x'(0) = Q \tag{6.17}$$

Note that the coefficients in the differential equation are all constants.

If we suppose that the solution $x(t)$ is a smooth function, then property (6.5) of the transform implies

$$\pounds\{x'\} = sX(s) - P,$$

$$\pounds\{x''\} = s^2X(s) - sP - Q.$$

Then

$$A(s^2X(s) - sP - Q) + B(sX(s) - P) + CX(s) = F(s); \text{ thus}$$

$$(As^2 + Bs + C)X(s) = APs + (AQ + BP) + F(s).$$

We can solve this equation easily for X(s) to obtain

$$X(s) = G(s) + H(s),$$

where

$$G(s) = \frac{APs + (AQ + BP)}{As^2+Bs+C}$$

$$H(s) = \frac{1}{As^2+Bs+C}F(s) = Y(s)F(s).$$

The solution x(t) to the initial value problem will have been found if we can find the inverse Laplace transforms g(t) and h(t). In the solved problems we present several examples illustrating that h(t) can be found from H(s) using Table 6.1 and the convolution formula, property (6.11). In addition, g(t) can be found from G(s) without having to resort to convolution.

In the event that the differential equation (6.17) models some physical system, we can think of the solution x(t) as consisting of two parts: g(t) represents the response of the system to the initial state, while h(t) represents the response of the system to the forcing term, f(t).

Transfer Functions

In analyzing the forced response, h(t), note that H(s) is in the form of a product, Y(s)F(s). The function

$$Y(s) = \frac{1}{As^2+Bs+C}$$

is referred to as the *transfer function* for the differential equation (6.17). The inverse transform of Y(s) is often referred to in the engineering literature as the *weight function.* In particular, since F(s)=1 if f(t)=δ(t), the weight function can be interpreted as the *impulse resonse;* i.e., y(t) represents the response of the system to a unit impulse forcing term. Once this response is known, the response of the system to any other forcing function f(t) is just equal to the convolution of f(t) with the impulse response, y(t).

Example 6.4 Solving an Initial Value Problem

6.4(A) CONSIDER THE INITIAL VALUE PROBLEM FOR Y=Y(t),

$$y'(t) - 3y(t) = 1, \; y(0) = 10.$$

If we let $Y(s) = \pounds\{y(t)\}$ then $\pounds\{y'(t)\} = sY(s) - 10$ and

$$sY(s) - 10 - 3Y(s) = 1/s.$$

We solve this equation for Y(s) to obtain

$$Y(s) = \frac{10}{s-3} + \frac{1}{s}\frac{1}{s-3} = Y_H(s) + Y_P(s).$$

According to entry 2 in Table 6.1,

$$£^{-1}\{Y_H(s)\} = 10\, e^{3t}.$$

Here we use $£^{-1}$ to indicate the inverse Laplace transform operation. This operation is accomplished here by looking in Table 6.1.

In order to find the inverse Laplace transform of $Y_P(s)$, we first use the method of *partial fractions* to write

$$\frac{1}{s}\frac{1}{s-3} = \frac{A}{s} + \frac{B}{s-3} = \frac{A(s-3)+Bs}{s(s-3)}$$

In order for this string of equalities to be valid, it is necessary that

$$A(s-3) + Bs = 1 \text{ for all s.}$$

In particular, for s=3 this implies 3B=1 and for s=0 it implies –3A=1; thus A=–1/3, B=1/3 and

$$\frac{1}{s}\frac{1}{s-3} = \frac{1}{3}\frac{1}{s-3} - \frac{1}{3}\frac{1}{s}.$$

Now, from Table 6.1 we have,

$$£^{-1}\{Y_P(s)\} = 1/3e^{3t} - 1/3.$$

6.4(B)

Consider the initial value problem,

$$x'(t) + tx(t) = f(t),\ x(0) = 0.$$

Note that the equation here is linear but has a variable coefficient. If we let X(s) denote the Laplace transform of x(t), then

$$£\{x'(t)\} = sX(s) - 0,$$

$$£\{tx(t)\} = -X'(s),$$

and the initial value problem for x(t) transforms into the following problem for X(s):

$$sX(s) - X'(s) = F(s)$$

This is again a differential equation. While the Laplace transform is an effective tool for solving linear equations with *constant* coefficients, it is generally not very helpful for equations with *variable* coefficients.

SOLVED PROBLEMS

Properties of the Laplace Transform

PROBLEM 6.1

Verify property (6.4) for the Laplace transform.

SOLUTION 6.1

Suppose f(t) and f′(t) are of exponential type and continuous for $t \geq 0$ (it is sufficient that f′ is piecewise continuous). Then, by the definition of the Laplace transform,

$$\pounds\{f'\} = \int_0^\infty f'(t)\, e^{-st}\, dt\,.$$

Integrating by parts leads to

$$\begin{aligned}\pounds\{f'\} &= f(t)e^{-st}\Big|_0^\infty - \int_0^\infty f(t)(-se^{-st})dt\,,\\ &= 0 - f(0) + s\,F(s).\end{aligned}$$

Here we used the fact f(t) is of exponential type to infer that $f(t)e^{-st}$ tends to zero as t tends to infinity. Now, applying this result to the second derivative f″(t) leads to

$$\begin{aligned}\pounds\{\,f''\,\} &= s\,\pounds\{\,f'\,\} - f'(0)\\ &= s\,[\,s\,F(s) - f(0)\,] - f'(0),\\ &= s^2F(s) - sf(0) - f'(0).\end{aligned}$$

This process may be applied repeatedly in order to obtain the transforms of derivatives of higher order.

PROBLEM 6.2

Verify that the Laplace transform has properties (6.8) and (6.9).

SOLUTION 6.2

Suppose f(t) is piecewise continous and of exponential type. Then the same is true of the product exp[bt]f(t) for any real number b; then

$$\pounds\{e^{bt}f(t)\} = \int_0^\infty f(t)e^{bt}e^{-st}dt$$

$$= \int_0^\infty f(t)e^{-(s-b)t}dt = F(s-b).$$

Similarly, for b positive,

$$\pounds\{H(t-b)f(t-b)\} = \int_b^\infty f(t-b)e^{-st}dt$$

$$= \int_0^\infty f(\tau)e^{-s(\tau+b)}d\tau = e^{-bs}F(s).$$

PROBLEM 6.3

Show that if f(t) has a Laplace transform F(s) then

$$\pounds\left\{\int_0^t f(\tau)d\tau\right\} = F(s)/s.$$

SOLUTION 6.3

Let

$$g(t) = \int_0^t f(\tau)d\tau\ .$$

Then $g'(t) = f(t)$ and $g(0)=0$. It follows now from property (6.5) that

$$\pounds\{g'\} = F(s) = s\,G(s) - 0;$$

thus $G(s) = F(s)/s$.

Laplace Transform Pairs

PROBLEM 6.4

Find the Laplace transform of the following functions:

(a) Cosh at and Sinh at

(b) tSinh at

(c) $f(t) = \begin{cases} \text{Sin } t & \text{if } 0 < t < \pi \\ 0 & \text{if } t > \pi \end{cases}$

(d) $f(t) = \begin{cases} 0 & \text{if } t < 4 \\ t^2 & \text{if } t > 4 \end{cases}$

SOLUTION 6.4

Note that each of these functions is piecewise continuous and of exponential type.

6.4(a) $\mathcal{L}\{\text{Cosh at}\} = \mathcal{L}\{\frac{1}{2}(e^{at}+e^{-at})\}$

$$= \tfrac{1}{2}(1/(s-a) + 1/(s+a))$$

$$= s/(s^2-a^2)$$

Now using the result of the previous problem, with g(t)=Sinh at and f(t)= aCosh at, we have

$$\mathcal{L}\{\text{Sinh at}\} = a/(s^2-a^2).$$

6.4(b) $\mathcal{L}\{\text{tSinh at}\} = -d/ds(a/(s^2-a^2))$

$$= 2as/(s^2-a^2)^2$$

6.4(c) Note that since Sin t = –Sin(t–π),

$$f(t) = (1 - H(t-\pi))\text{Sin } t$$

$$= \text{Sin } t + \text{Sin}(t-\pi)\, H(t-\pi)$$

Then we can apply property (6.9):

$$F(s) = \frac{1}{s^2+1} + \frac{e^{-\pi s}}{s^2+1} = \frac{1+e^{-\pi s}}{s^2+1}$$

6.4(d) Here $f(t) = t^2H(t-4)$. In order to apply property (6.9), we must rewrite this in the form

$$f(t) = (t-4+4)^2H(t-4)$$

$$= (t-4)^2H(t-4) + 8(t-4)H(t-4) + 16H(t-4).$$

Then we have

$$F(s) = 2\,\frac{e^{-4s}}{s^3} + 8\,\frac{e^{-4s}}{s^2} + 16\,\frac{e^{-4s}}{s}$$

PROBLEM 6.5

Find the Laplace transform of the following functions:

(a) $f(t) = \begin{cases} t & \text{if } 0 < t < T \\ 0 & \text{if } t > T. \end{cases} \qquad T > 0.$

(b) $f_k(t) = \begin{cases} t-kT & \text{if } kT < t < (k+1)T \\ 0 & \text{otherwise} \end{cases} \qquad k=0,1,\ldots$

(c) $g(t) = \sum_{k=0}^{\infty} f_k(t)$

SOLUTION 6.5

6.5(a) The function f(t) is a "ramp function"; it can be rewritten

$$f(t) = t - t\,H(t-T).$$

Then

$$F(s) = \frac{1}{s^2} - \frac{e^{-sT}}{s^2}.$$

6.5(b) For k=0,1,... $f_k(t)$ is just the function f(t) from part (a), shifted to the right by the amount kT. Then

$$\begin{aligned} f_k(t) &= (t-kT)(H(t-kT) - H(t-(k+1)T)) \\ &= (t-kT)\,H(t-kT) - (t-(k+1)T)\,H(t-(k+1)T) \\ &\qquad - T\,H(t-(k+1)T) \end{aligned}$$

and

$$F_k(s) = \frac{e^{-skT}}{s^2} - \frac{e^{-s(k+1)T}}{s^2} - \frac{T}{s}e^{-s(k+1)T}.$$

6.5(c) Finally in g(t) a series of ramp functions combine to form a so-called "sawtooth" wave. We have

$$G(s) = \sum_{k=0}^{\infty} F_k(s) = \frac{1 - e^{-sT}(1 + sT)}{s^2} \sum_{k=0}^{\infty} e^{-skT}$$

Then the formula for the sum of a geometric series leads to the result

$$G(s) = \frac{1-e^{-sT}(1+sT)}{s^2(1-e^{-sT})} = \frac{1}{s^2} - \frac{T}{s}\frac{e^{-sT}}{1-e^{-sT}}$$

Inverse Transforms

PROBLEM 6.6

Find the inverse Laplace transform for each of the following:

(a) $1/(s-2)$

(b) $s/(s^2-3s+2)$

(c) $1/[s(s^2-3s+2)]$

(d) $1/[s(s^2+1)]$

(e) $e^{-3s}/(s-1)^2$

(f) $1/(s^2+1)^2$

SOLUTION 6.6

6.6(a) According to entry 2 in table 6.1,

$$\pounds^{-1}\{1/(s-2)\} = e^{2t}.$$

6.6(b) Since, $s^2 - 3s + 2 = (s-2)(s-3)$ we can use partial fractions to write

$$\frac{s}{(s-2)(s-3)} = \frac{A}{s-2} + \frac{B}{s-3} = \frac{A(s-3) + B(s-2)}{(s-2)(s-3)} .$$

This is an identity in s if A = –2 and B = 3. Then

$$\pounds^{-1}\{\frac{s}{(s-2)(s-3)}\} = \pounds^{-1}\{\frac{3}{s-3}\} - \pounds^{-1}\{\frac{2}{s-2}\}$$

$$= 3e^{3t} - 2e^{2t}.$$

6.6(c) We can use partial fractions to write

$$\frac{1}{s(s-2)(s-3)} = \frac{A}{s} + \frac{B}{s-2} + \frac{C}{s-3}$$

$$= \frac{A(s-2)(s-3) + Bs(s-3) + Cs(s-2)}{s(s-2)(s-3)}$$

This leads to A = 1/6, B = –1/2, C = 1/3, and

$$\pounds^{-1}\{\frac{1}{s(s-2)(s-3)}\} = \frac{1}{6} - \frac{e^{2t}}{2} + \frac{e^{3t}}{3} .$$

6.6(d) Here we can use partial fractions or we can use the result of problem 6.3. We have from table 6.1 that,

$$\pounds^{-1}\{1/(s^2+1)\} = \text{Sin } t.$$

Then from problem 6.3, we have

$$\pounds^{-1}\{1/(s(s^2+1))\} = \int_0^t \text{Sin } \tau \, d\tau = 1 - \text{Cos } t.$$

6.6(e) This inverse can be computed in three steps. First, from Table 6.1

$$\pounds^{-1}\{1/s^2\} = t\,;$$

then, using property (6.8),

$$\pounds^{-1}\{1/(s-1)^2\} = te^t;$$

finally, by property (6.9),

$$\pounds^{-1}\{e^{-3s}/(s-1)^2\} = H(t-3)(t-3)e^{t-3}.$$

6.6(f) Write

$$\frac{1}{(s^2+1)^2} = \frac{1}{2s}\,\frac{2s}{(s^2+1)^2} = \frac{-1}{2s}\,\frac{d}{ds}\,\frac{1}{s^s+1}$$

Now

$$\pounds^{-1}\{-d/ds(1/(s^2+1))\} = t\text{Sin } t,$$

and hence

$$\pounds^{-1}\{1/(s^2+1)^2\} = \tfrac{1}{2}\int_0^t \tau \text{ Sin } \tau \, d\tau$$

$$= \tfrac{1}{2}\,(\text{Sin } t - t\text{Cos } t).$$

PROBLEM 6.7

Let

$$f(t) = \begin{cases} t & \text{if } 0 < t < 1 \\ 2-t & \text{if } 1 < t < 2 \\ 0 & \text{if } t > 2 \end{cases}$$

Then find:

(a) $\pounds^{-1}\{sF(s)\}$

(b) $\mathcal{L}^{-1}\{e^{-2s}sF(s)\}$

(c) $\mathcal{L}^{-1}\{sF(s-1)\}$

SOLUTION 6.7

6.7(a) By property (6.5),

$$\mathcal{L}^{-1}\{sF(s)\} = f'(t) + f(0).$$

But f(0)=0 and

$$f'(t) = \begin{cases} 1 & \text{if } 0 < t < 1 \\ -1 & \text{if } 1 < t < 2 \\ 0 & \text{if } t > 2 \end{cases}$$

$$= H(t) - 2H(t-1) + H(t-2); \text{ so}$$

$$\mathcal{L}^{-1}\{sF(s)\} = H(t) - 2H(t-1) + H(t-2).$$

6.7(b) Using this result, along with property (6.9), leads to

$$\mathcal{L}^{-1}\{e^{-2s}sF(s)\} = H(t-2) - 2H(t-3) + H(t-4).$$

6.7(c) First, by property (6.8),

$$\mathcal{L}^{-1}\{F(s-1)\} = e^{t}f(t) = g(t).$$

Then g(0)=0 and

$$\mathcal{L}^{-1}\{sF(s-1)\} = g'(t) + g(0)$$

$$= e^{t}(f(t)+f'(t)).$$

Applications of the Transform

PROBLEM 6.8

The current, i(t), in an RL-circuit satisfies

$$Li'(t) + Ri(t) = f(t), \ t > 0.$$

Solve for i(t) if:

(a) $i(0) = 0$ and $f(t) = \text{Sin}\ \Omega t$.

(b) $i(0) = 0$ and $f(t) = 1 - H(t-T)$ for $T > 0$.

SOLUTION 6.8

6.8(a) If I(s) and F(s) denote the Laplace transforms of i(t) and f(t) respectively, then

$$LsI(s) - 0 + RI(s) = F(s);$$

hence

$$I(s) = \frac{1}{L}\frac{F(s)}{s+k}, \; k = R/L$$

Here

$$F(s) = \frac{\Omega}{s^2 + \Omega^2}$$

and (by partial fractions)

$$\frac{1}{s+k}\,\frac{1}{s^2+\Omega^2} = \frac{1}{s+k} + \frac{Bs+c}{s^2+\Omega^2}$$

for

$$A = 1/(k^2+ \Omega^2)$$

$$B = -1/(k^2+ \Omega^2)$$

$$C = k/(k^2+ \Omega^2).$$

Then

$$i(t) = (A\Omega/L)(e^{-kt} - \text{Cos } \Omega t + (k/\Omega)\text{Sin } \Omega t).$$

6.8(b) For the conditions of part (b)

$$F(s) = (1 - e^{-sT})/s$$

and

$$I(s) = \frac{1}{L}\frac{1-e^{-Ts}}{s(s+k)}$$

Then

$$i(t) = 1/R(1 - e^{-kt} - H(t{-}T)(1 - e^{-k(t-T)})).$$

PROBLEM 6.9

Consider the following equation for a damped spring-mass system:

$$m\,x''(t) = -Kx(t) - Cx'(t) + f(t),\ t > 0.$$

Solve for x(t) if:

(a) $f(t) = 0,\ x(0) = A,\ x'(0) = B.$

(b) $f(t) = \delta(t),\ x(0) = x'(0) = 0.$

(c) $f(t) = H(t),\ x(0) = x'(0) = 0.$

SOLUTION 6.9

6.9(a) If we denote the Laplace transform of x(t) by X(s), then

$$m(s^2X(s) - As - B) = -KX(s) - C(sX(s) - A);\ \text{so}$$
$$(ms^2 + Cs + K)X(s) = mAs + mB + CA.$$

Note that $ms^2+Cs+K = P(s)$ is the characteristic polynomial that we have encountered previously in connection with second-order linear equations with constant coefficients. Let us write,

$$\begin{aligned} P(s) &= m(s^2 + (C/m)s + C^2/4m^2 + (K/m - C^2/4m^2)) \\ &= m((s+\sigma)^2 + \Omega^2), \quad \text{if } C^2 < 4mK \\ &= m((s+\sigma)^2 + 0) \quad \text{if } C^2 = 4mK \\ &= m((s+\sigma)^2 - \Omega^2) \text{ if } C^2 > 4mK \end{aligned}$$

where

$$\sigma = C/(2m),\ \Omega^2 = |K/m - C^2/4m^2|.$$

Then since

$$P(s)X(s) = mA(s+\sigma) + m(B + A\sigma),$$

we find

$$X(s) = A\frac{s+\sigma}{(s+\sigma)^2+\Omega^2} + \frac{B+A\sigma}{\Omega}\frac{\Omega}{(s+\sigma)^2+\Omega^2} \quad (C^2<4mK)$$

$$= \frac{A}{s+\sigma} + \frac{B+A\sigma}{(s+\sigma)^2} \quad (C^2=4mK)$$

$$= \frac{A(s+\sigma) + B+A\sigma}{(s+\sigma-\Omega)(s+\sigma+\Omega)} = \frac{D_1}{s+\sigma-\Omega} + \frac{D_2}{s+\sigma-\Omega} \quad (C^2>4mK)$$

In the case ($C^2>4mK$), the parameters D_1 and D_2 can be found in terms of A, B, and σ by means of partial fractions. Then

$$x(t) = Ae^{-\sigma t}\text{Cos }\Omega t + (B+A\sigma)/\Omega\ e^{-\sigma t}\text{Sin }\Omega t \qquad (C^2 < 4mK)$$

$$= Ae^{-\sigma t} + (B+A\sigma)te^{-\sigma t} \qquad (C^2 = 4mK)$$

$$= D_1\exp(-(\sigma+\Omega)t) + D_2\exp(-(\sigma-\Omega)t) \qquad (C^2 > 4mK)$$

These are the damped, critically damped, and overdamped responses of the spring-mass system, discussed previously in Chapter 5. These are all transient responses. In the case that C=0, corresponding to zero damping, we have $\sigma = 0$, and the solution becomes

$$x(t) = A\text{Cos }\Omega t + (B/\Omega)\text{Sin }\Omega t. \qquad (C=0)$$

Thus, as the damping parameter increases from zero, the solution changes character. It begins at C=0 as a periodic response and changes to an oscillating solution with decaying amplitude as C^2 varies in the interval $0 < C^2 < 4mK$. At $C^2=4mK$ the solution ceases all oscillation and simply decays steadily to zero with increasing time.

6.9(b) In this case $x(0) = x'(0) = 0$, and we say that the system is *relaxed.* The forcing term is the unit impulse and the solution is the so called impulse response. The Laplace transform of the impulse response is the *transfer function,* which we shall denote by Y(s). Then

$$P(s)Y(s) = 1\ ;$$

i.e.,

$$Y(s) = \frac{1}{P(s)}$$

$$= \frac{1}{m\Omega}\ \frac{\Omega}{(s+\sigma)^2+\Omega^2} \qquad \text{if } 0 \le C^2 < 4mK$$

$$= \frac{1}{m}\ \frac{1}{(s+\sigma)^2} \qquad \text{if } C^2 = 4mK$$

$$= \frac{1}{m}\ \frac{1}{(s+\sigma-\Omega)(s+\sigma+\Omega)} \qquad \text{if } C^2 > 4mK$$

The inverse transform of Y(s) is the *weight function;* we shall denote it by W(t). For this system the character of the weight function varies with C:

$$W(t) = (1/m\Omega)e^{-\sigma t}\text{Sin }\Omega t \qquad \text{if } 0\ C^2 < 4mK$$

$$= (1/m)te^{-\sigma t} \qquad \text{if } C^2=4mK$$

$$= (D_1/m)\exp(-(\sigma-\Omega)t) + (D_2/m)\exp(-(\sigma+\Omega)t)) \qquad \text{if } C^2>4mK$$

To compute the response, x(t), of the relaxed system to any other forcing term, f(t), note that

$$P(s)X(s) = F(s);$$

i.e.,

$$X(s) = Y(s)\,F(s).$$

Then x(t) is obtained by using property (6.11) of the Laplace transform:

$$x(t) = \int_0^t W(t-\tau)f(\tau)d\tau.$$

6.9(c) In particular, when f(t) = H(t), we refer to x(t) as the *step response.* Then

$$x(t) = \int_0^t W(t-\tau)\,d\tau.$$

Thus, the step response is the antiderivative of the impulse response (the impulse response is the derivative of the step response).

PROBLEM 6.10

Consider the forced, undamped spring-mass system,

$$m\,x''(t) = -Kx(t) + f(t),\ t > 0,$$

in the following cases:

(a) $x(0) = x'(0) = 0$

$f(t) = A \text{ Sin } pt$ where p^2 does not equal $\Omega^2 = K/m$

(b) $x(0) = x'(0) = 0$

$f(t) = A \text{ Sin } \Omega t$ where $\Omega^2 = K/m$.

SOLUTION 6.10

6.10(a) Proceding as in the last problem, we obtain

$$X(s) = \frac{A}{m}\,\frac{1}{s^2+\Omega^2}\,\frac{p}{s^2+p^2}$$

$$= \frac{A}{m}\,\frac{p}{p^2-\Omega^2}\left[\frac{1}{s^2+\Omega^2} - \frac{1}{s^2+p^2}\right]$$

We used partial fractions here to go from the first line to the second. Then, inverting the transform:

$$x(t) = \frac{A}{m} \frac{1}{p^2-\Omega^2} ((p/\Omega)\text{Sin } \Omega t - \text{Sin } pt) .$$

This is a periodic function whose amplitude varies periodically. See also Problem 5.9(a).

6.10(b) In case the frequency of the forcing term is equal to Ω, the so called natural frequency of the system, we obtain

$$x(s) = \frac{A}{m} \frac{\Omega}{(s^2 + \Omega^2)^2}$$

This inverse must be computed like part (f) of Problem 6.6. Then

$$x(t) = \frac{A\Omega}{2m} \int_0^t \tau \text{ Sin } \Omega\tau \, d\tau$$

$$= (A/2m)(\text{Sin } \Omega t - (1/\Omega) \, t\text{Cos } \Omega t).$$

This is an oscillatory function whose amplitude grows steadily with increasing time due to the tCos Ωt term. The phenomenon of a bounded input causing an unbounded response is called *resonance*. Resonance occurs when the frequency of the periodic forcing term equals the natural frequency of the system. In the presence of damping (i.e. $C > 0$) the resonant response is much larger than the response when the driving frequency is not equal to the natural frequency, but it is not unbounded. See Problems 5.9(b) and 5.10.

The Laplace transform provides a systematic way of transforming a linear differential equation with constant coefficients in an unknown function x(t) into algebraic equation in the transform X(s). It is a particularly effective way of solving inhomogeneous differential equations.

The TRANSFER FUNCTION is defined as Y(s) = 1/P(s), where P(s) denotes the characteristic polynomial of the differential equation. Then the inverse Laplace transform of Y(s) may be interpreted as the response of the relaxed system to a unit impulse input. The response of the relaxed system to any other forcing term, f(t), is then simply the convolution product of f(t) with the impulse response.

The Laplace transform is not generally suitable for nonlinear equations nor equations with variable coefficients. Tables 6.1 and 6.2 summarize the properties of the transform and list a sufficient number of transform pairs to permit the application of the method in a large number of examples.

SUPPLEMENTARY PROBLEMS

Find f(t) if F(s) is given:

6.1 $F(s) = 3/s - 1/(s+1) + 2/(s^2+1)$

6.2 $F(s) = 4/(s^2+16) - 8/(s^2-4s+20)$

6.3 $F(s) = 2/(3s-2)^2$

6.4 $F(s) = 5/(s^3-s^2-s-1)$

6.5 $F(s) = 5/(s^2+2s+17)$

6.6 $F(s) = 1/(s^2-s+2)$

6.7 $F(s) = 6/(s-a)^4$

6.8 $F(s) = 1/(s+1)^2$

6.9 $F(s) = e^{-sT}/s^2$

6.10 $F(s) = (s^2-a^2)/(s^2+a^2)^2$

Find the Laplace transform of the following:

6.11 $f(t) = \text{Cosh } 9t$

6.12 $f(t) = t \text{ Sinh } 4t$

6.13 $f(t) = (t+1)^2$

6.14 $f(t) = (t-2)e^t$

6.15 $f(t) = \begin{cases} 1 & \text{if } 1<t<2 \\ -2 & \text{if } 2<t<5 \\ 0 & \text{if } t>5 \end{cases}$

Solve by using the Laplace transform

6.16 $y'' + y' - 12y = 0$ $y(0) = 1,$ $y'(0) = 0.$

6.17 $y'' + y' - 12y = 0$ $y(0) = 0,$ $y'(0) = 1.$

6.18 $y'' + y' - 12y = 1$ $y(0) = 0,$ $y'(0) = 0.$

6.19 $(D^4-1)y(t) = t$ $y(0)=y'(0)=y''(0)=y^{(3)}(0)=0.$

6.20 $y' + y = e^{-t}$ $y(0) = 1.$

6.21 $3y^{(3)} + 5y'' + y' - y = 0$, $y(0)=0$, $y'(0)=1$, $y''(0)= -1.$

Use the Laplace transform to evaluate the convolution products:

6.22 $t*t$

6.23 $e^t*\text{Sin}t$

6.24 $1*1$

6.25 $H(t-T)*f(t)$

ANSWERS TO SUPPLEMENTARY PROBLEMS

6.1 $f(t) = 3 - e^{-t} + 2\text{Sin } t$

6.2 $f(t) = \text{Sin } 4t - 2e^{2t}\text{Sin } 4t$

6.3 $f(t) = 2te^{2t/3}/9$

6.4 $f(t) = 5(2t\, e^t - e^t + e^{-t})/4$

6.5 $f(t) = (5e^{-t}\text{Sin } 4t)/4$

6.6 $f(t) = 2\, 7^{-1/2}\, e^{t/2}\, \text{Sin}(7^{-1/2}t/2)$

6.7 $f(t) = t^3e^{at}$

6.8 $f(t) = te^{-t}$

6.9 $f(t) = (t-T)H(t-T)$

6.10 $f(t) = t\text{Cos } at$

6.11 $F(s) = s/(s^2-81)$

6.12 $F(s) = 8s/[(s-4)^2(s+4)^2]$

6.13 $F(s) = 2/s^3 + 2/s^2 + 1/s$

6.14 $F(s) = 1/(s-1)^2 - 2/(s-1)$

6.15 $F(s) = e^{-s}/s - e^{-2s}/s - 2e^{-3s}/s + 2e^{-5s}/s$

6.16 $y(t) = (3e^{-4t} + 4e^{3t})/7$

6.17 $y(t) = (e^{3t} - e^{-4t})/7$

6.18 $y(t) = (e^{-4t} + 4e^{3t}/3)/28 - 1/12$

6.19 $y(t) = \text{Cosh } t - \text{Cos } t - t^2$

6.20 $y(t) = (t+1)e^{-t}$

6.21 $y(t) = (9\,e^{t/3} + (4t-9)e^{-t})/16$

6.22 $t^3/6$

6.23 $(\Omega e^t - \sin \Omega t - \Omega \cos \Omega t)/(1+\Omega^2)$

6.24 t

6.25 $H(t-T) \int_0^{t-T} f(\tau)\, d\tau$

7

Systems of Linear Equations

In each of the previous chapters we have considered problems involving a single differential equation for one unknown function. We considered equations of various orders and types (e.g., linear, nonlinear, constant coefficient, variable coefficient, etc.), but always there was just one equation in a single unknown function. In this chapter we shall consider systems of *simultaneous linear differential equations. Such systems arise naturally in a number of applications and, in addition, any linear differential equation of order N can be converted to a system of N equations of order one.*

Recall that in order to obtain a unique solution for a system of simultaneous linear algebraic equations, the number of equations must equal the number of unknowns. We shall find that a similar requirement applies to systems of linear differential equations. Other principles from linear algebra, intended originally for dealing with systems of linear algebraic equations, have applications to systems of linear differential equations.

We shall show how to construct solutions to systems of equations having constant coefficients. When the number of equations is small, say two or three equations, the Laplace transform is an effective tool for constructing solutions. When the number of equations is large, then more sophisticated techniques are required. These techniques rely heavily on ideas from linear algebra.

INTRODUCTION TO SYSTEMS

Physical Problems Leading to Systems of Equations

Systems of differential equations arise naturally out of a variety of physical applications.

ELECTRICAL NETWORKS

In Chapter 5 we considered simple electrical circuits consisting of electrical components (i.e. , resistors, capacitors, inductances, and electromotive forces) connected in series. These circuits were governed by a single differential equation in which the current was the unknown function. This worked because a series circuit consists of a single closed loop, and so the same current flows through every component. In circuits consisting of two or more connected loops the situation is more complicated. We shall refer to such circuits as *networks.* A point in a network from which two or more wires run to different components will be called a *node.* The following general principles apply to any network:

Kirchhoff's Voltage Law—The algebraic sum of the voltage drops around any simple closed loop equals zero.

Kirchhoff's Current Law—The algebraic sum of the currents flowing to any node equals zero.

In addition to these general principles we have empirical laws for the voltage drop across each type of component. These are given in Chapter 5 by equations (5.3), (5.4), and (5.6). Applying Kirchhoff's laws together with the empirical equations for the voltage drops to any given network leads to a system of differential equations for the unknown currents in the network. Several examples are given in the solved problems.

MECHANICAL SYSTEMS WITH SEVERAL DEGREES OF FREEDOM

In Chapter 5 we showed that the behavior of a spring-mass system is governed by a single differential equation in which the displacement of the mass from its equilibrium position is the unknown function of time. Such a system is said to have a single *degree of freedom.* In a more complicated system we may have several masses connected to one another by springs. The displacement of each mass is then an unknown function of time. By applying Newton's laws we may be able to derive a system of differential equations that must be satisfied by the displacement functions. Examples of systems with several degrees of freedom are found in the solved problems.

PROBLEMS REDUCIBLE TO SYSTEMS

In addition to such physical applications, systems can arise in other ways. For example, every linear equation of order N in one unknown function is equivalent to a system of N linear equations of order one involving N unknown functions. The two problems are equivalent in the sense that a solution of either of them leads directly to a solution for the other. It is also true that a system of N simultaneous equations of order two is equivalent to a system of 2N equations of order one.

Example 7.1 Problems Reducable to First Order Systems

7.1(A)

Consider the following initial value problem of order N for the unknown function y=y(t):

$$y^{(N)}(t) + a_{N-1}\, y^{(N-1)}(t) + \dots + a_1\, y'(t) + a_o\, y(t) = f(t) \tag{7.1}$$

$$y(0)=b_o,\ y'(0) = b_1, \dots, y^{(N-1)} = b_{N-1} \tag{7.2}$$

We now define N new unknown functions as follows:

$$u_1(t)=y(t),\ u_2(t)=y'(t), \dots, u_N(t)=y^{(N-1)}(t) \tag{7.3}$$

Then these new unknown functions satisfy the following system of linear equations:

$$u_1'(t) = u_2(t)$$

$$u_2'(t) = u_3(t)$$

$$\cdot$$

$$\cdot$$

$$\cdot$$

$$u_N'(t) = -a_{N-1}\, u_N - \dots - a_o u_1 + f(t)\,.$$

These equations may be expressed in matrix notation as

$$d/dt\ U(t) = AU(t) + F, \tag{7.4}$$

where $U(t) = [u_1, \dots, u_N]^T$, $F = [0, \dots, 0, f]^T$ are column vectors and A denotes the following N by N matrix of coefficients:

$$A = \begin{bmatrix} 0 & 1 & 0 & \dots & 0 \\ 0 & 0 & 1 & \dots & 0 \\ & \dots & & & 1 \\ -a_o & -a_1 & & & -a_{N-1} \end{bmatrix}$$

Matrix notation and other notions from linear algebra are reviewed in a later section of this chapter.

The initial value problem (7.1), (7.2) for a single equation of order N has been converted to the initial value problem consisting of the system of N first order equations (7.4) together with the initial condition,

$$U(0) = [b_o, \ldots, b_{N-1}]^T \tag{7.5}$$

Once (7.4), (7.5) has been solved, the solution for (7.1), (7.2) can be obtained from (7.3). We shall discuss methods for solving systems like (7.4) later in the chapter.

7.1(B)

Consider the following system of two equations of order two for the two unknown functions $y_1 = y_1(t)$, $y_2 = y_2(t)$

$$\begin{aligned} y_1'' &= a_{11}y_1 + a_{12}y_2 + b_{11}y_1' + b_{12}y_2' + f_1 \\ y_2'' &= a_{21}y_1 + a_{22}y_2 + b_{21}y_1' + b_{22}y_2' + f_1; \end{aligned} \tag{7.6}$$

in matrix notation:

$$Y'' = A\,Y + B\,Y' + F,$$

where

$$Y = [y_1, y_2]^T, F = [f_1, f_2]^T,$$

and

$$A = \begin{bmatrix} a_{11} & a_{12} \\ a_{21} & a_{22} \end{bmatrix} \qquad B = \begin{bmatrix} b_{11} & b_{12} \\ b_{21} & b_{22} \end{bmatrix}$$

If we define new unknown functions,

$$u_1 = y_1,\ u_2 = y_2,\ u_3 = y_1',\ u_4 = y_2',$$

then

$$\begin{aligned} u_1' &= u_3 \\ u_2' &= u_4 \\ u_3' &= a_{11}u_1 + a_{12}u_2 + b_{11}u_3 + b_{12}u_4 + f_1 \\ u_4' &= a_{21}u_1 + a_{22}u_2 + b_{21}u_3 + b_{22}u_4 + f_2\,; \end{aligned}$$

i.e.,

$$
\frac{d}{dt}\begin{bmatrix} u_1 \\ u_2 \\ u_3 \\ u_4 \end{bmatrix} = \begin{bmatrix} 0 & 0 & 1 & 0 \\ 0 & 0 & 0 & 1 \\ a_{11} & a_{12} & b_{11} & b_{12} \\ a_{21} & a_{22} & b_{21} & b_{22} \end{bmatrix}\begin{bmatrix} u_1 \\ u_2 \\ u_3 \\ u_4 \end{bmatrix}\begin{bmatrix} 0 \\ 0 \\ f_1 \\ f_2 \end{bmatrix} \tag{7.7}
$$

The first-order system (7.7) is equivalent to the second order system (7.6). While it is not necessary to convert a second order system to a system of order one for purposes of solution, it is convenient for purposes of stating theorems of existence and uniqueness since a single theorem, stated for a system of order one, then covers linear systems of all orders.

Theorem 7.1 Existence of a Solution for a First-order System

Theorem 7.1 Consider the system of N linear equations in the N unknown functions, $x_1(t), \ldots, x_N(t)$:

$$
\begin{aligned}
x_1'(t) &= a_{11}(t)x_1(t) + \ldots + a_{1N}(t)x_N(t) + f_1(t) \\
&\;\;\vdots \\
x_N'(t) &= a_{N1}(t)x_1(t) + \ldots + a_{NN}(t)x_N(t) + f_N(t)
\end{aligned} \tag{7.8}
$$

If each of the functions $a_{jk}(t)$, $j,k=1, \ldots, k$ and $f_k(t)$, $k=1,\ldots,N$ is continuous on the interval $T_o \leq t \leq T_1$ then for each t_o in $[T_o,T_1]$ and each set of N constants $C_o,\ldots,C_N$ there exists a unique set of functions $x=x_k(t)$ $k=1,\ldots,N$ such that

$$
\begin{aligned}
&\text{(a) } x_1(t),\ldots,x_N(t) \text{ solves (7.8)} \\
&\text{(b) } x_1(t_o)=C_1, \ldots, xN(t_o)=C_N .
\end{aligned} \tag{7.9}
$$

Theorem 7.1 plays the same role for systems of equations that Theorem 4.3 plays for single equations. In fact, since single equations can be transformed to first order systems, Theorem 7.1 implies Theorem 4.3. Note also that since second order systems can be transformed to equivalent first-order systems, it is not necessary to have a separate existence theorem for systems of order two. Finally, the theorem implies that solutions for (7.8) can always be found and that when initial conditions are imposed, there is exactly one solution. However, the theorem does not suggest how solutions may be constructed. When the coefficients a_{jk} are variable, there is no general algorithm for constructing solutions, but for the case of constant coefficients there are various techniques available.

APPLICATIONS OF THE LAPLACE TRANSFORM TO SYSTEMS

We saw in Chapter 6 that the Laplace transform changes a constant coefficient linear differential equation into an algebraic problem in which the unknown is the Laplace transform of the solution to the differential equation. We can apply this technique to systems of linear differential equations as well, provided that the coefficients are constants. The technique is most effective when the number of equations is small.

Example 7.2 A Simple Application of the Laplace Transform

Consider the following initial value problem for a system of two equations in two unknown functions, $x_1(t)$ and $x_2(t)$:

$$x_1'(t) = 2x_1(t) - 5x_2(t) \qquad x_1(0)=7$$

$$x_2'(t) = x_1(t) - 2x_2(t), \qquad x_2(0)=3.$$

If we let $X_1(s)$ and $X_2(s)$ denote the Laplace transforms of the two unknown functions, then it follows from results derived in Chapter 6 that

$$sX_1 - 7 = 2X_1 - 5X_2$$

$$sX_2 - 3 = X_1 - 2X_2 .$$

Thus,

$$\begin{bmatrix} s-2 & 5 \\ -1 & s+2 \end{bmatrix} \begin{bmatrix} X_1 \\ X_2 \end{bmatrix} = \begin{bmatrix} 7 \\ 3 \end{bmatrix}$$

We can solve this system of two algebraic equations by Cramer's rule or any other method that is convenient. We obtain

$$X_1(s) = \frac{7s-1}{s^2+1} \text{ and } X_2(s) = \frac{3s+1}{s^2+1} .$$

It follows from entries 4 and 5 in Table 6.1 that

$$x_1(t) = 7\text{Cos}t - \text{Sin}t \text{ and } x_2(t) = 3\text{Cos}t + \text{Sin}t.$$

Additional applications of the Laplace transform may be found in the solved problems. Note that the method is only effective so long as the system of algebraic equations for the transforms of the unknowns can be easily solved. When the number of unknowns is large, it may be more efficient to apply techniques from linear algebra.

ELEMENTARY LINEAR ALGEBRA

Matrix Notation

Analysis of systems of linear differential equations requires at least a rudimentary knowledge of linear algebra. Matrix notation and the terminology of linear algebra provide a framework in which it is possible to efficiently discuss a variety of mathematical problems, including systems of algebraic equations, matrix eigenvalue problems, and problems in differential equations. Here we shall briefly review a few essentials.

VECTORS AND SCALARS

We shall be concerned with objects we shall call vectors and another type of object we shall call scalars, together with two algebraic operations, *vector addition* and *scalar multiplication.* We shall take the complex numbers as our collection of scalars, and we shall define a vector to mean an N-tuple of scalars. We use the notation X to denote a column vector of scalars and the notation X^T to represent a row vector. That is,

$$X = \begin{bmatrix} x_1 \\ \cdot \\ \cdot \\ \cdot \\ x_N \end{bmatrix} \quad \text{and } X^T = [x^1, ..., x^N] .$$

The need for distinction between a row and a column vector will become apparent when we consider matrix multiplication.

We define the operations of vector addition and scalar multiplication by

$$X^T \pm Y^T = [x_1 \pm y_1, ..., x_N \pm y_N] \text{ for all vectors } X \text{ and } Y$$

$$\alpha X^T = [\alpha x_1, ..., \alpha x_N] \text{ for all vectors } X \text{ and scalars } \alpha.$$

INNER PRODUCT AND ORTHOGONALITY

In addition, we define the *inner* product (or dot product, as it is sometimes called) of two vectors to mean,

$$X \cdot Y = \sum_{k=1}^{N} x_k \bar{y}_k ,$$

where $\bar{y}_k$ denotes the complex conjugate of the number y_k. Two vectors X and Y are said to be *orthogonal* if their inner product equals zero. We use the notation $X \perp Y$ to indicate that $X \cdot Y = 0$ i.e. that X and Y are orthogonal.

MATRICES AND MATRIX OPERATIONS

We shall define an M by N *matrix* to mean an array of numbers consisting of M rows with N entries in each row. Equivalently, it is an array of N columns with M entries in each column. In particular, then, a vector X is an N by 1 matrix and a row vector X^T is a 1 by N matrix.

The operations of addition and scalar multiplication for matrices are defined entrywise by

$$[A] \pm [B] = [a_{jk} \pm b_{jk}] \text{ for } [A] = [a_{jk}] \text{ and } [B] = [b_{jk}]$$

$$\alpha[A] = [\alpha a_{jk}] \text{ for } [A] = [a_{jk}] \text{ and scalar } \alpha .$$

Note that addition and subtraction are defined only for two matrices of the same dimensions. One cannot form the sum or difference of an M by N matrix with another matrix that is not of dimension M by N.

We shall also define multiplication for matrices. The product of the M by N matrix A times the N by K matrix B is the M by K matrix C whose (i,j) entry is the number

$$c_{ij} = \sum_{k=1}^{N} a_{ik} b_{kj} \text{ for } i=1,...,M \text{ and } j=1,...,K .$$

Note that the matrix product AB is defined only if the number of columns of A is equal to the number of rows of B and that the product matrix, C, then has the same number of rows as A and the same number of columns as B. Matrix multiplication has the following properties:

i) AB is not in general equal to BA

ii) A(B+C)= AB+AC

iii) A(BC) = (AB)C

MATRIX INVERSES

The N by N matrix I whose entries satisfy,

$$I = \begin{cases} 1 \text{ if } j=k \\ 0 \text{ if } j \text{ is not equal to } k \end{cases}$$

is called the *identity matrix.* The identity matrix satisfies, IA = AI = A for all N by N matrices A. An N by N matrix A is said to be *invertible* if there exists another N by N matrix B such that AB = BA = I. In this case we write $B=A^{-1}$ and say that B is the *inverse* of A. Only a square matrix can have an inverse but not every square matrix A has an inverse. One way of distinguishing a matrix which has an inverse from a matrix for which no inverse exists is by means of the determinant.

THE DETERMINANT

The *determinant* of an N by N matrix A is defined by

$$\text{if } N=2, \det A = a_{11}a_{22} - a_{12}a_{21}$$

$$\text{if } N>2, \det A = \sum_{k=1}^{N} (-1)^{j+k} a_{jk} M_{jk}.$$

Here M_{jk} denotes the determinant of the N–1 by N–1 matrix obtained by deleting from A the jth row and kth column.

CRAMER'S RULE

Note that the definition of det A for N>2 is independent of the choice of k, $1 \le k \le N$. We can show that an N by N matrix A has an inverse if and only if det A is different from zero. In addition, we have the following result known as *Cramer's Rule:*

> if $AX = b$ and det A is different from zero, then the entries of the solution vector X are given by
>
> $$x_k = \det A(k)/\det A \text{ for } k=1,\ldots,N,$$
>
> where A(k) denotes the N by N matrix obtained by replacing the kth column of A with the vector b.

FORMULAS FOR THE MATRIX PRODUCT

If A denotes an N by N matrix and X denotes an N by 1 matrix (i.e., a vector), then the product AX is defined and produces an N by 1 matrix (vector). This product can be expressed in two equivalent ways:

$$AX = \begin{bmatrix} a_{11}x_1+\ldots+a_{1N}x_N \\ \cdot \\ \cdot \\ \cdot \\ a_{N1}x_1+\ldots+a_{NN}x_N \end{bmatrix} = \begin{bmatrix} R_1 \cdot X \\ \cdot \\ \cdot \\ \cdot \\ R_N \cdot X \end{bmatrix} \tag{7.10}$$

and

$$AX = x_1C_1 + \ldots + x_NC_N, \tag{7.11}$$

where $R_1,\ldots,R_N$ denote the N rows of the matrix A and $C_1,\ldots,C_N$ denote the columns. In particular, the system of linear algebraic equations,

$$a_{11}x_1 + \ldots + a_{1N}x_N = b_1$$
$$\vdots$$
$$a_{N1}x_1 + \ldots + a_{NN}x_N = b_N$$

can be expressed in matrix notation by writing $AX=b$. Similarly, the system of linear differential equations (7.8) is written in matrix notation as

$$d/dt X(t) = AX+F(t).$$

HERMITIAN TRANSPOSE OF A MATRIX

Note also that for any N by N matrix A and N-vectors X and Y, we have

$$(AX)\cdot Y = X\cdot(A^*Y), \tag{7.12}$$

where A^* is the *Hermitian transpose* of the matrix A; i.e. $A^* = [\mathring{a}_{kj}]$ and $A = [a_{jk}]$. Here $\mathring{a}_{kj}$ indicates the complex conjugate of a_{kj}. We can show that for N by N matrices A and B,

$$(AB)^* = B^*\,A^*. \tag{7.13}$$

An N by N matrix A with the property that $A=A^*$ is said to be a *Hermitian* matrix.

LINEAR ALGEBRA TERMINOLOGY

The collection of all possible N-tuples is an example of a *vector space.* It will be convenient to be able to refer to special collections of vectors in a vector space.

Definition 7.1 A collection M of vectors is said to form a *subspace* if, for all vectors X and Y in M and all scalars α and β, the *linear combination,* $\alpha X + \beta Y$ also belongs to M.

Example 7.3 Subspaces

7.3(A)

Let A denote an N by N matrix and consider the collection of all N-tuples X that satisfy the condition, $AX = 0$. We refer to this collection as the *null space* of the matrix A and denote it by NS[A]. To see that this collection forms a subspace, note that if X and Y denote arbitrary vectors in the null space, (i.e. $AX = 0$ and $AY = 0$) then for any scalars α and β,

$$A(\alpha X + \beta Y) = A(\alpha X) + A(\beta Y)$$
$$= \alpha AX + \beta AY = 0\,;$$

i.e. the linear combination $\alpha X + \beta Y$ belongs to NS[A] for arbitrary scalars α and β.

7.3(B)

Let $V_1, \dots, V_p$ denote a set of p distinct vectors, and consider the collection of all possible linear combinations of these vectors. We refer to this collection as the *span* of the vectors $V_1, \dots, V_p$, and we denote it by writing, span$[V_1, \dots, V_p]$. It is clear from the definition that this collection of vectors is a subspace. We refer to the vectors $V_1, \dots, V_p$ as a *spanning* set for this subspace.

7.3(C)

Let M denote a collection of vectors, and consider the collection consisting of all vectors V that are orthogonal to each of the vectors in M. We denote this collection by $M^\perp$. That is, V belongs to $M^\perp$ if and only if $V \cdot X = 0$ for every X in M. Then we can show that $M^\perp$ is a subspace (whether or not the collection M is a subspace). We refer to $M^\perp$ as the *orthogonal complement* of M.

Definition 7.2 The vectors $V_1, \dots, V_p$ are *linearly dependent* if there exist scalars $\alpha_1, \dots, \alpha_p$, not all equal to zero, such that

$$\alpha_1 V_1 + \dots + \alpha_p V_p = 0. \tag{7.14}$$

If the set of vectors is not linearly dependent then it is *linearly independent;* i.e., the V's are linearly independent if (7.14) implies $\alpha_1 = \alpha_2 = \dots = \alpha_p = 0$.

Let X denote a vector in the subspace, span$[V_1, \dots, V_p]$. By the definition of this subspace there exist scalars $\alpha_1, \dots, \alpha_p$ such that $\alpha_1 V_1 + \dots + \alpha_p V_p = X$. In general there may exist many choices for these scalars but when the vectors $V_1, \dots, V_p$ are linearly independent, then for each X in the subspace there is a unique set of scalars; i.e. X can be written as a linear combination of the V's in exactly one way. This motivates the following definition.

Definition 7.3 The vectors $V_1, \dots, V_p$ form a *BASIS* for the subspace M if:

(a) M = span$[V_1, \dots, V_p]$ and

(b) the vectors $V_1, \dots, V_p$ are linearly independent.

In this case we say that the subspace M has *dimension* equal to p.

Example 7.4 Bases

7.4(A)

For k=1,...,N let e_k denote the N-tuple having a 1 as its kth entry and every other entry equal to zero. Then it is evident that every N-tuple of scalars can be written as a linear combination of the vectors $e_1, \dots, e_N$; i.e., the vectors e_1 through e_N form a spanning set for the vector space of N-tuples. In addition, it is not difficult to prove that the vectors e_1 through e_N are linearly independent. Then according to Definition 7.3, these vectors form a basis for the vector space of N-tuples of scalars. We often denote the vector space of N-tuples by C^N. It follows from Definition 7.3 that the dimension of C^N is equal to N. We refer to the vectors e_k as the *standard basis* for C^N.

7.4(B)

The vectors $e_1 = [1,0]^T$ and $e_2 = [0,1]^T$ form a basis for C^2. Note also that the vectors $E_1 = [1,1]^T$ and $E_2 = [1,-1]^T$ form a spanning set for C^2; i.e. $e_1 = (E_1 + E_2)/2$ and $e_2 = (E_1 - E_2)/2$ so any vector written as a linear combination of e_1 and e_2 can also be written as a linear combination of E_1 and E_2. In addition, it is not hard to show that E_1 and E_2 are linearly independent. Thus E_1 and E_2 are also a basis for C^2. We may find other, different bases for C^2 (infinitely many of them), but each of them will contain precisely two vectors. More generally, every basis for C^N contains N vectors, and if M is a subspace of C^N having a basis consisting of p vectors, then every basis for M contains p vectors.

Definition 7.4 A basis $[u_1,...,u_p]$ with the property that

$$u_j \cdot u_k = \begin{cases} 1 \text{ if } j=k \\ 0 \text{ if } j \text{ differs from } k \end{cases}$$

is said to be an *orthonormal* basis.

The computational convenience of an orthonormal basis will be illustrated in the solved problems. We now state (without proof) several results that will be important for what follows.

Important Results

Relevant to the solution of systems of simultaneous equations, we have the following theorem.

Theorem 7.2

Theorem 7.2 Let A denote an N by N matrix.

(a) The system of equations $AX = b$ has a unique solution X for every N-tuple b if and only if det A is different from zero.

(b) The homogeneous system $AX = 0$ has nontrivial solutions, (i.e. solutions other than $X = 0$) if and only if det A = 0.

The following statements are equivalent (i.e. each implies and is implied by all of the others):

1. $AX = b$ has a unique solution for every b in C^N.

2. Det A is different from zero.

3. The dimension of NS[A] equals zero.

4. The rows of A are linearly independent as vectors.

5. The columns of A are linearly independent as vectors.

6. A^{-1} exists.

EIGENVALUES AND EIGENVECTORS

For an N by N matrix A, consider the problem of finding nonzero vectors X and scalars λ such that $AX = \lambda X$. Clearly, $X = 0$ solves the equation for all choices of the scalar λ but for certain choices of λ, called *eigenvalues,* there exist nonzero solution vectors X. These nonzero solution vectors are called the *eigenvectors* corresponding to the eigenvalue λ. Note that if X is an eigenvector for A corresponding to the eigenvalue zero, then αX is also an eigenvector corresponding to the eigenvalue λ for all choices of the scalar α. In particular, we can always multiply an eigenvector by a scaling factor so that it has unit length. An eigenvector of unit length is a *normalized eigenvector.*

Theorem 7.3

Theorem 7.3 Let A denote an N by N matrix. Then the following are equivalent:

1. λ is an eigenvalue for A.
2. $(A - \lambda I)X = 0$ for nonzero vector X.
3. $\text{Det}(A - \lambda I) = 0$.

If λ is an eigenvalue for A, then X is an eigenvector for A, corresponding to the eigenvalue λ, if and only if X belongs to $NS[A - \lambda I]$.

It is easy to show that if A is an N by N matrix, then $\det(A - \lambda I) = 0$ is a polynomial equation of degree N in the unknown λ. Since every such equation has at least one root, it follows that every N by N matrix has at least one eigenvalue and corresponding eigenvector. More precisely, we have the following theorem.

Theorem 7.4

Theorem 7.4 If the N by N matrix A has N *distinct* eigenvalues, the corresponding eigenvectors form a basis for C^N. If the N by N matrix A is Hermitian, then

(a) the eigenvalues of A are all real numbers;

(b) eigenvectors corresponding to distinct eigenvalues are orthogonal;

i.e., $NS[A - \lambda I] \perp NS[A - \mu I]$ for μ not equal to λ; and

(c) the normalized eigenvectors of A form an orthonormal basis for C^N.

Suppose that the eigenvectors of the N by N matrix A form a basis for C^N, and let P denote the N by N matrix whose columns are these N eigenvectors, $V_1,\ldots,V_N$. Then the matrix product AP is defined and, in fact,

$$AP = PD\,; \tag{7.15}$$

i.e.,

$$AP = A[V_1,\dots,V_N] = [AV_1,\dots,AV_N]$$
$$= [\lambda V_1,\dots,\lambda V_N] = PD,$$

where D denotes the N by N matrix whose diagonal entries are the eigenvalues of A and whose off-diagonal entries are zeroes. That is,

$$d_{jk} = \begin{cases} \lambda_j \text{ if } j=k \\ 0 \text{ if } j \text{ does not equal } k. \end{cases}$$

Example 7.5 Eigenvalue Problems

7.5(A)

To find the eigenvalues and eigenvectors of the 2 by 2 matrix

$$A = \begin{bmatrix} 2 & 3 \\ 1 & 4 \end{bmatrix}.$$

we compute det $(A - \lambda I) = (2 - \lambda)(4 - \lambda) - 3 = \lambda^2 - 6\lambda + 5$. It follows easily that the eigenvalues of A, the roots of $\det(A - \lambda I) = 0$, are $\lambda = 1$ and $\lambda = 5$.

To find the eigenvector corresponding to $\lambda = 1$, we find the nontrivial solutions of $(A - I)X = 0$; i.e., the system

$$\begin{bmatrix} 1 & 3 \\ 1 & 3 \end{bmatrix}\begin{bmatrix} x_1 \\ x_2 \end{bmatrix} = \begin{bmatrix} 0 \\ 0 \end{bmatrix}$$

has solutions, $x_1 = -3x_2$. Then $X_{(1)}^T = [-3x_2,x_2]^T = x_2[-3,1]^T$ is, for all choices of the scalar x_2, an eigenvector for A corresponding to the eigenvalue $\lambda = 1$. In a similar way, we solve $(A - 5I)X = 0$ to find that $X_{(5)}^T = [x_2,x_2]^T = x_2[1,1]^T$ is, for all choices of the scalar x_2, an eigenvector for A corresponding to the eigenvalue $\lambda = 5$.

7.5(B)

It is considerably more difficult to find the eigenvalues and eigenvectors of the N by N Hermitian matrix A,

$$A = \begin{bmatrix} -2 & 1 & 0 & \dots & 0 \\ 1 & -2 & 1 & \dots & 0 \\ & & & \dots & \\ & & & & 1 \\ 0 & & & 1 & -2 \end{bmatrix} \tag{7.16}$$

By writing out each of the equations in the system $(A - \lambda I)X = 0$, it is easy to see that the entries of an eigenvector $X = [x_1, \dots ,x_N]^T$ must satisfy

$$x_{k-1} - (2+\lambda)x_k + x_{k+1} = 0 \text{ for } k=1,\ldots,N$$

where (7.17)

$$x_o = x_{N+1} = 0.$$

In solved Problem 7.13, we solve the problem (7.17) and show that the matrix A from (7.16) has the N real eigenvalues

$$\lambda_n = 2[\text{Cos}(n\pi/(N+1)) - 1] \; n = 1,\ldots,N \tag{7.18}$$

and that corresponding to λ_n we have the eigenvector,

$$X_n = [\text{Sin}(n\pi/(N+1)), \text{Sin}(2n\pi/(N+1)), \ldots, \text{Sin}(Nn\pi/(N+1))]^T \tag{7.19}$$

SYSTEMS OF LINEAR DIFFERENTIAL EQUATIONS

The system of linear differential equations of the form (7.8) can be solved by means of the Laplace transform if the coefficients are constant and if N is small, say 2 or 3. For large systems, however, it may be more efficient to apply techniques from linear algebra. In particular, the following theorem states that when the eigenvectors of the coefficient matrix A form a basis for C^N then the solution of the associated initial value problem can be written in terms of the eigenvalues and eigenvectors of A.

Theorem 7.5

Theorem 7.5 Suppose the N by N matrix A has N independent eigenvectors X_1 to X_N corresponding to eigenvalues λ_1 to λ_N. Then for each given initial vector Γ, the unique solution of the initial value problem,

$$d/dt\,U(t) = AU(t) \qquad U(0) = \Gamma \tag{7.20}$$

is given by,

$$U(t) = C_1 e^{\lambda_1 t} X_1 + \ldots + C_N e^{\lambda_N t} X_N \tag{7.21}$$

where $C = [C_1, \ldots, C_N]^T$ is obtained by solving the system $PC = \Gamma$ where P denotes the matrix whose columns are the eigenvectors of A. For each forcing function $F(t)$, the unique solution of the initial value problem,

$$d/dt\, U(t) = AU(t) + F(t), \qquad U(0) = 0 \tag{7.22}$$

is given by,

$$U(t) = \int_0^t e^{\lambda_1(t-\tau)} G_1(\tau)\, d\tau\, X_1 + \ldots + \int_0^t e^{\lambda_N(t-\tau)} G_N(\tau)\, d\tau\, X_N \tag{7.23}$$

where $G(t) = [G_1(t), \ldots, G_N(t)]^T$ is obtained by solving the system $PG = F$. In particular, if the eigenvectors form an orthonormal basis, then $G_k = F \cdot X_k$ for $k = 1,\ldots,N$.

Note In problem 7.21 we justify the use of the notation

$$U(t) = e^{At}\Gamma \tag{7.24}$$

in place of (7.21) and , in place of (7.23),

$$U(t) = \int_0^t e^{A(t-\tau)} F(\tau)\, d\tau\,. \tag{7.25}$$

SOLVED PROBLEMS

Mathematical Models Leading to Systems

PROBLEM 7.1

Consider the circuit shown in Figure 7.1. Find a system of differential equations satisfied by the currents in this network.

SOLUTION 7.1

The electromotive force E(t) induces currents $i_1(t)$, $i_2(t)$, and $i_3(t)$ in the circuit. According to Kirchhoff's voltage law the algebraic sum of

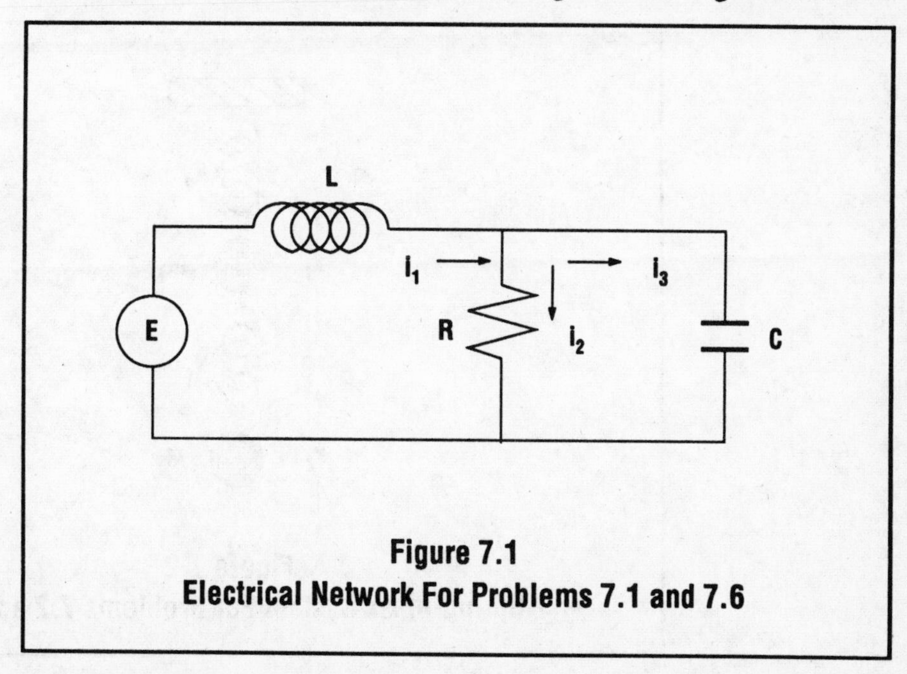

Figure 7.1
Electrical Network For Problems 7.1 and 7.6

the voltages around each of the two loops of this circuit equals zero. That is,

$$E(t) - Li_1'(t) - Ri_2(t) = 0 \tag{1}$$

$$1/C \int^t i_3 - Ri_2(t) = 0\,. \tag{2}$$

In addition, Kirchhoff's current law states that the algebraic sum of currents flowing to any node must equal zero. For this circuit, that implies

$$i_1 - i_2 - i_3 = 0. \tag{3}$$

Now, differentiating equation (2) with respect to t and using this, together with (1) and (3), we obtain

$$Li_1' + Ri_2 = E(t)$$

$$Ri_2' - 1/C\,(i_1 - i_2) = 0\,.$$

This is a system of two differential equations for the unknown currents i_1 and i_2. Once these are known, the current i_3 can be obtained via (3). See Problem 7.6 for a solution.

PROBLEM 7.2

Consider the system of two masses and two springs shown in Figure 7.2. Derive a system of differential equations that is satisfied by the displacements of the masses from their equilibrium positions.

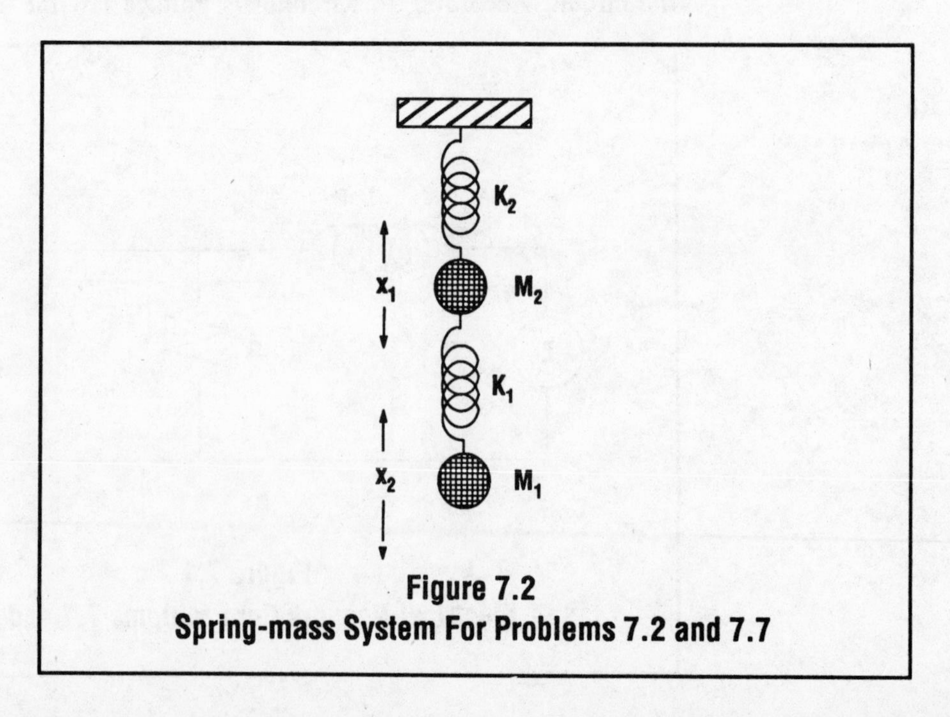

Figure 7.2
Spring-mass System For Problems 7.2 and 7.7

SOLUTION 7.2

If we let $x_1(t)$ and $x_2(t)$ denote the displacements of masses m_1 and m_2, respectively, then spring number one (with stiffness equal to K_1) is stretched by the amount $(x_1 - x_2)$ while the other spring is stretched by the amount x_2. Then Hooke's law implies that

$$-K(x_1 - x_2) = \text{restoring force exerted on } m_1 \text{ by spring 1}$$

$$K(x_1 - x_2) = \text{restoring force exerted on } m_2 \text{ by spring 1}$$

$$-K_2x_2 = \text{restoring force exerted on } m_2 \text{ by spring 2.}$$

Then Newton's law states that $F = ma$; i.e.,

$$m_1x_1''(t) = -K_1(x_1 - x_2)$$

$$m_2x_2''(t) = K_1(x_1 - x_2) - K_2x_2 .$$

This is a system of two second-order differential equations for the displacements of the two masses from their equilibrium positions.

PROBLEM 7.3

Consider a thin, straight tube of uniform cross section that has been filled with water. We inject a contaminating substance into the water and this contaminant then moves through the tube by the process of molecular diffusion. Show that this process can be modeled by a system of ordinary differential equations. Assume that the initial distribution of contaminant is known and that the two ends of the tube are in contact with reservoirs in which the level of contaminant is specified.

SOLUTION 7.3

We suppose that the walls of the tube are impervious to the flow of fluid and the width of the tube is small compared to its length. Then the flow of contaminant can be assumed to be 1-dimensional; i.e. the concentration of contaminant varies only with the position *along* the tube and with time.

The tube is assumed to lie with its centerline along the x-axis with its left end located at $x=0$ and its right end at $x=L$. Here L denotes the length of the tube. We think of the tube as composed of a number of identical "cells." If the number of cells is denoted by N, then each of the cells is of width $w= L/N$ and each is assumed to have cross sectional area A (FIGURE 7.3).

For $k=1,\ldots,N$ we let

$$u_k(t) = \text{concentration of contaminant in cell } k \text{ at time } t.$$

Then the *amount* of contaminant in cell k at time t is equal to $wAu_k(t)$, and it follows that

$$d/dt\,[wAu_k(t)] = \text{net inflow to cell } k. \tag{1}$$

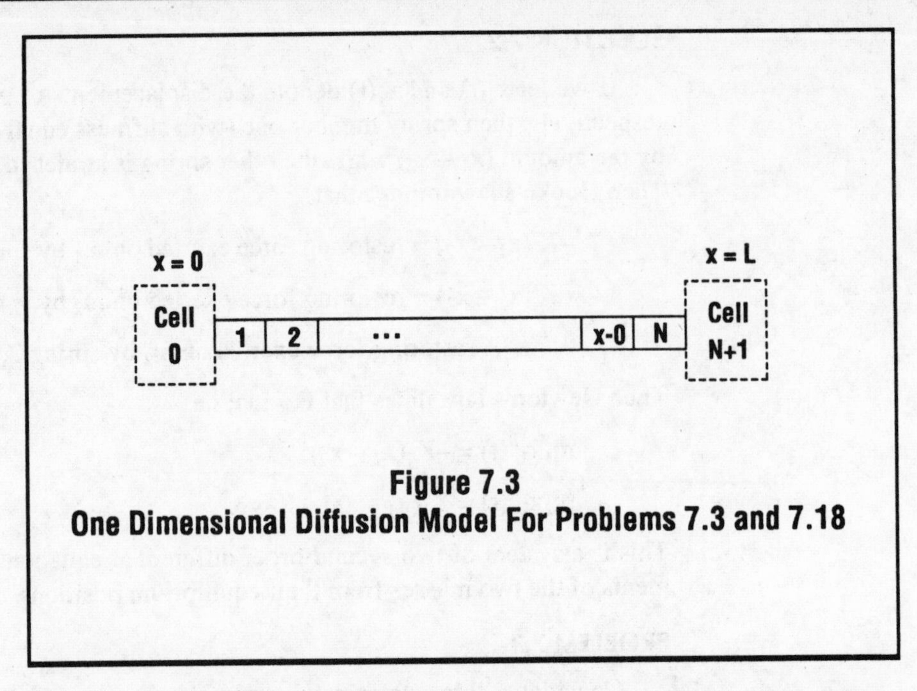

Figure 7.3
One Dimensional Diffusion Model For Problems 7.3 and 7.18

We are assuming that contaminant is neither created nor destroyed within any cell, hence the rate of change of amount of contaminant in the cell is due only to inflow or outflow across the two end faces of the cell. An empirical principle known as *Fick's law* states that diffusion proceeds such that contaminant flows from regions of high concentration to regions of low concentration at a rate that is proportional to the "concentration gradient." In our notation, this is expressed by

$$\text{inflow rate across left face of cell k} = A\,D\,\frac{u_{k-1} - u_k}{w} \tag{2}$$

$$\text{outflow rate across right face of cell k} = A\,D\,\frac{u_k - u_{k+1}}{w} \tag{3}$$

Here D denotes a positive, material dependent parameter called the *diffusivity.* Large values of D correspond to rapidly diffusing materials. According to (2), if the concentration in cell k–1 is higher than the concentration in cell k, then the flow across the left face of cell k has a positive value, corresponding to an inflow. A similar observation can be made with regard to (3) and the flow across the face at the right side of cell k. The difference (2) minus (3) represents the net inflow rate to cell k. Combining (1), (2) and (3) leads to

$$d/dt\; wAu_k(t) = AD(u_{k-1} - 2u_k + u_{k+1})/w\,;$$

that is, for each k=1,2,...,N

$$d/dt[u_k(t)] = D(u_{k-1} - 2u_k + u_{k+1})/w^2. \tag{4}$$

We are assuming here that the ends of the tube are in contact with reservoirs in which the concentration of contaminant is specified. If we label the concentration in the reservoirs at the left and right ends of the tube as u_o and u_{N+1}, respectively, then we assume that $u_o(t)$ and $u_{N+1}(t)$ are known functions of t. For purposes of this example, let us suppose that each of these functions equals zero. In addition we suppose that the *initial* concentration in each cell is known; i.e., $u_k(0)$ is known for k=1, ... ,N. Then applying (4) in each of the N cells in the tube leads to the following system of equations:

$$u_1'(t) = -2\vartheta u_1 + \vartheta u_2 \qquad u_1(0)=C_1$$

$$u_2'(t) = \vartheta u_1 - 2\vartheta u_2 + \vartheta u_3 \qquad u_2(0)=C_2$$

$$\vdots$$

$$u_N'(t) = \vartheta u_{N-1} - 2\vartheta u_N \qquad u_N(0)=C_N.$$

Here we let $\vartheta = D/w^2$. This system can be written in matrix notation as

$$d/dt\ U(t) = \vartheta B\ U(t), \qquad U(0)=C, \tag{5}$$

where $U(t) = [u_1,...,u_N]^T$, denotes the vector of unknown functions, $C = [C_1,...,C_N]^T$, denotes the initial data vector, and B denotes the N by N coefficient matrix,

$$B = \begin{bmatrix} -2 & 1 & 0 & \cdots & 0 \\ 1 & -2 & 1 & & 0 \\ & & & \cdots & \\ & & & & 1 \\ 0 & & & 1 & -2 \end{bmatrix} \tag{6}$$

The solution of the system (5) is discussed in Problem 7.18.

PROBLEM 7.4

Consider the one-dimensional diffusion model described in the previous problem. Suppose, however, that the ends of the tube are sealed so that no contaminant flows out the ends of the tube. Write down the system of equations that models the diffusion process in this case.

SOLUTION 7.4

In the previous problem the ends of the tube were open to reservoirs in which the concentration of contaminant was specified. Here we wish to model the situation in which the ends of the tube are sealed. We say that the two problems have different *boundary conditions.*

Using the notation and terminology of the previous problem, we see that if the flow across the left end of the tube (i.e., the left face of cell 1) is zero, then the only flow in or out of cell 1 is the outflow across the right face of that cell. Then

$$d/dt[wAu_1(t)] = -DA\,\frac{u_1(t) - u_2(t)}{w} \tag{1}$$

Similarly, if the right end of the tube (i.e., the right face of cell N) is sealed, then the only flow into or out of cell N is the inflow at the left face. This leads to

$$d/dt\; wA\,u_N(t) = DA\,\frac{u_{N-1}(t) - u_N(t)}{w} \tag{2}$$

In cells 2 through N–1 the balance equation is the same as it was in the last problem; i.e., equation (4). Thus, we have the following system of equations for the unknown concentrations:

$$u_1'(t) = -\vartheta u_1 + \vartheta u_2 \qquad u_1(0)=C_1$$

$$u_2'(t) = \vartheta u_1 - 2\vartheta u_2 + \vartheta u_3 \qquad u_2(0)=C_2$$

$$\vdots$$

$$u_N'(t) = \vartheta u_{N-1} - \vartheta u_N \qquad u_N(0)=C_N.$$

That is

$$d/dt\; U(t) = \vartheta B'\,U(t), \qquad U(0)=C, \tag{4}$$

where $U(t)$ and C are as in the previous problem, but with the coefficient matrix changed to

$$B' = \begin{bmatrix} -1 & 1 & 0 & \cdots & 0 \\ 1 & -2 & 1 & & \\ & & & \cdots & \\ & & & -2 & 1 \\ 0 & & & 1 & -1 \end{bmatrix} \tag{5}$$

Hence, the change in the boundary conditions has the effect of slightly changing the coefficient matrix B in the system of differential equations. We shall see in Problem 7.18 what the corresponding effect is on the solution to the system.

PROBLEM 7.5

Consider a system of N identical masses connected to one another and to a pair of rigid supports by N+1 springs as shown in Figure 7.4. Let $x_k(t)$ denote the displacement of the kth mass from its equilibrium position. Then show that these displacements satisfy

$$d^2/dt^2\, X(t) = \mu\, B\, X(t) \tag{1}$$

where $\mu = K/m$ and B denotes the N by N matrix given by (6) in Problem 7.3. Here m and K denote the common values of mass and spring stiffness for the system.

SOLUTION 7.5

The only forces acting on each mass are the elastic spring forces exerted by the springs attached to its left and right sides. For the kth mass these forces are equal to

$$F_{left} = -K[x_k(t) - x_{k-1}(t)] \tag{2}$$

$$F_{right} = -K[x_{k+1} - x_k(t)]. \tag{3}$$

Then Newton's second law implies

$$\begin{aligned} m\, d^2/dt^2\, x_k(t) &= F_{left} - F_{right} \\ &= K(x_{k-1} - 2x_k + x_{k+1}) \end{aligned} \tag{4}$$

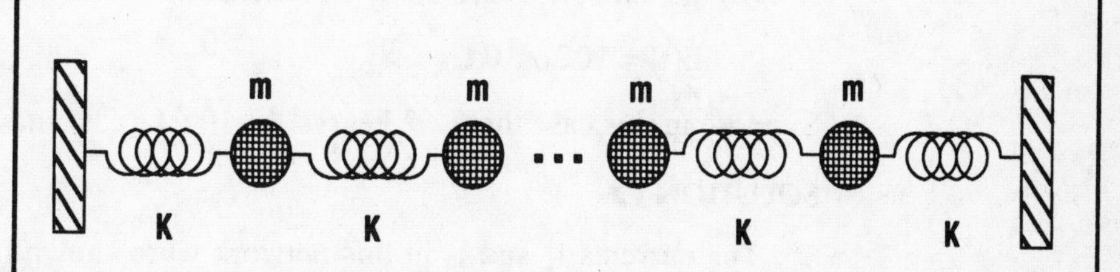

Figure 7.4
Spring-mass System With N Identical Springs and Masses For Problem 7.5

for k=1,2, ..., N. Our assumptions on the rigid supports at the right and left of the spring-mass system may be interpreted to mean

$$x_o = \text{the displacement of the left support} = 0,$$
$$x_{N+1} = \text{the displacement of the right support} = 0. \qquad (5)$$

Then applying (4) for k=1,...,N and using the conditions (5) leads to

$$d^2/dt^2\, x_1(t) = -2\,\mu\, x_1 + \mu\, x_2$$
$$d^2/dt^2\, x_2(t) = \mu\, x_1 - 2\mu\, x_2 + \mu\, x_3$$
$$\vdots$$
$$d^2/dt^2\, x_N(t) = \mu\, x_{N-1} - 2\mu\, x_N\,.$$

Expressing this system in matrix notation yields (1). Note that in order to solve (1) we shall need two initial conditions,

$$X(0) = C \text{ and } d/dt\, X(0) = D\,,$$

where C and D denote N-tuples of specified constants. The system (1) can be used to model a variety of physical systems, including the longitudinal vibrations of a homogeneous elastic shaft and acoustic waves in a pipe.

Application of the Laplace Transform to Systems

PROBLEM 7.6

Consider the network in Figure 7.1. Find the current i_3 through the capacitor if the network is initially at rest and,

(a) the electromotive force is produced by a battery; i.e.,

$$E(t) = E_o \text{ for } E_o \text{ a constant.}$$

(b) the network is driven by an alternating current; i.e. ,

$$E(t) = 10\text{Cos}10\, t.$$

Suppose in this case that L=2 henrys, $C=10^{-2}$ farads, and R=10 ohms.

SOLUTION 7.6

The currents i_1 and i_2 in this network were shown (in Problem 7.1) to satisfy

$$Li_1' + Ri_2 = E(t)$$
$$Ri_2' - 1/C\,(i_1 - i_2) = 0\,.$$

If the network is initially at rest, then we may suppose that

$$i_1(0) = 0 \text{ and } i_2(0) = 0\,.$$

Then

$$sI_1(s) + R/L\, I_2(s) = 1/L\, E(s)$$
$$sI_2(s) - 1/RC\, I\,(s) + 1/RC\, I_2(s) = 0;$$

i.e.,

$$\begin{bmatrix} s & R/L \\ -1/RC & s + 1/RC \end{bmatrix} \begin{bmatrix} I_1 \\ I_2 \end{bmatrix} = \begin{bmatrix} 1/L\, E \\ 0 \end{bmatrix}$$

where I_1, I_2, and E denote the Laplace transforms of i_1, i_2, and E(t), respectively. We may solve for I_1 and I_2 and obtain

$$I_1(s) = \frac{E(s)}{L} \frac{s + 1/RC}{s(s + 1/RC) + 1/LC},$$

$$I_2(s) = \frac{E(s)}{L} \frac{1/RC}{s(s + 1/RC) + 1/LC},$$

hence,

$$I_3(s) = I_1 - I_2 = \frac{E(s)}{L} \frac{1/RC}{s(s + 1/RC) + 1/LC}.$$

Note that

$$s(s + 1/RC) + 1/LC = (s + 1/2RC)^2 + 1/LC - (1/2RC)^2$$
$$= (s + \beta)^2 + \omega^2,$$

where

$$\beta = 1/(2RC) \text{ and } \omega^2 = 1/LC - \beta^2 > 0.$$

(a) In the case that $E(t)=E_o$, we have $E(s) = E_o/s$ and

$$I_3(s) = \frac{E_o}{L} \frac{1}{(s + \beta)^2 + \omega^2}.$$

Then it follows from entry 5 of Table 6.2 and entry 4 of Table 6.1 that

$$i_3(t) = (E_o/\omega L)\, e^{-\beta t}\, \text{Sin}\omega t.$$

Thus, when the circuit is driven by a battery, the current through the capacitor is a decaying oscillatory function of time; i.e., the steady-state current i is zero.

(b) When the network is driven by an alternating current, we have

$$E(s) = \frac{10s}{s^2 + 100}$$

and

$$I_3(s) = \frac{5s}{s^2+100}\,\frac{s}{(s+5)^2+25}\,.$$

Here we used that for the given values of R,L, and C, β=5 and ω^2=25. We use partial fractions to rewrite $I_3(s)$ in the form

$$I_3(s) = \frac{2}{5}\,\frac{s+5}{s^2+100} - \frac{1}{5}\,\frac{2s+5}{(s+5)^2+25}$$

from which it follows that

$$i_3(t) = .8(2\,\mathrm{Cos}10t + \mathrm{Sin}10t) - .2\,e^{-5t}(2\mathrm{Cos}5t - \mathrm{Sin}5t).$$

In the case that the network is driven by a periodic forcing term, the current i_3 contains a transient term whose frequency is determined by the network components, but in the steady state the response is periodic with frequency equal to that of the forcing.

PROBLEM 7.7

Find the response of the spring-mass system of Problem 7.2 to the initial conditions

$$x_1(0)=a,\ x_2(0)=0, \quad \text{and} \quad d/dt\, x_2(0)=d/dt\, x_1(0)=0, \tag{1}$$

if the springs and masses are such that $K_2=3K_1$, $m_2=2m_1$, and

$$\Omega_1^{\,2}=K_1/m_1=64, \quad \Omega_2^{\,2}=K_2/m_2=96, \quad \Omega_o^{\,2}=K_1/m_2=32. \tag{2}$$

SOLUTION 7.7

In Problem 7.2 we showed that the displacements of the masses satisfy the equations

$$m_1x_1''(t) = -K_1(x_1-x_2)$$

$$m_2x_2''(t) = K_1(x_1-x_2) - K_2x_2\,.$$

Apply the Laplace transform in connection with the initial conditions (1) to obtain the following equations for the transforms of the unknown functions:

$$m_1[s^2X_1(s) - sa] = -K_1X_1 + K_1X_2$$

$$m_2[s^2X_2(s) - 0] = -(K_1 + K_2)X_2 + K_1X_1\,.$$

We can then solve these algebraic equations to obtain

$$X_1(s) = \frac{m_1\,s\,a\,(m_2s^2 + K_1 + K_2)}{(m_1s^2 + K_1)\,(m_2s^2 + K_1 + K_2) - K_1^{\,2}}$$

$$X_2(s) = \frac{K_1\, m_1\, s\, a}{(m_1 s^2 + K_1)\,(m_2 s^2 + K_1 + K_2) - K_1^2}$$

With the indicated values of the parameters, we have

$$(m_1 s + K_1)(m_2 s^2 + K_1 + K_2) - K_1^2 = m_1 m_2 (s^2 + 156)(s^2 + 36)$$

and

$$X_1(s) = a\, \frac{s(s^2 + 128)}{s^2 + 156)(s^2 + 36)} = a\,(\frac{7}{30}\frac{s}{s^2 + 156} + \frac{23}{30}\frac{s}{s^2 + 36})$$

$$X_2(s) = a\, \frac{32s}{(s^2 + 156)(s^2 + 36)} = a\, \frac{8}{30}\, (\frac{s}{s^2 + 156} - \frac{s}{s^2 + 36})$$

Here we used partial fractions to split up the expressions for $X_1(s)$ and $X_2(s)$. It follows then from entry 5 in Table 6.1 that

$$x_1(t) = a(7\text{Cos } 12.5t + 23 \text{ Cos } 6t)/30$$

$$x_2(t) = a(8\text{Cos } 12.5t - 8 \text{ Cos } 6t)/30$$

Note that both $x_1(t)$ and $x_2(t)$ exhibit "beats" (see Problem 5.9). That is, each of these displacements oscillates at a "fast frequency" with an amplitude that varies periodically at a "slow frequency."

PROBLEM 7.8

A transformer consists of a pair of electrical circuits that influence one another by means of "magnetic induction." Figure 7.5 shows a pair of such circuits. The circuit on the left is driven by an electromotive force $E(t) = E_o \text{Cos}\Omega t$, which then induces a current in the circuit on the right. Find the current induced in the right half of the circuit if

$L_1 = 2$ henrys,	$L_2 = 1$ henry,
$R_1 = 10$ ohms	$R_2 = 1$ ohm ,
$C = 10^{-1}$ farads	$\Omega = 10$ cps.

SOLUTION 7.8

If we let $i_1(t)$ and $i_2(t)$ denote the current in the left and right parts of the circuit, respectively, then it can shown that these functions satisfy

$$\begin{aligned} &L_1 i_1'(t) + M i_2'(t) + R_1 i_1(t) = E(t) \\ &L_2 i_2'(t) + M i_1'(t) + R_2 i_2(t) + 1/C \int^t i_2 = 0\,, \end{aligned} \qquad (1)$$

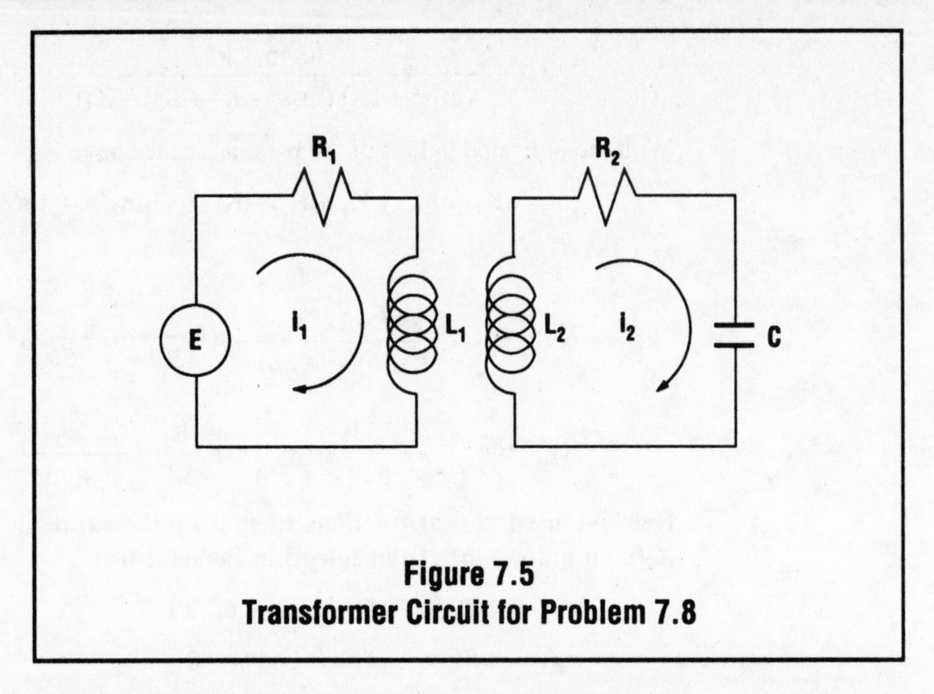

Figure 7.5
Transformer Circuit for Problem 7.8

where M denotes the so called "mutual inductance" between the two parts of the circuit. We suppose that $L_1L_2 > M^2$; in particular, let us assume that $M^2 = L_1L_2 - 1$; i.e., $M = 1$. We can express (1) equivalently in the form

$$\begin{aligned} &L_1 i_1''(t) + M i_2''(t) + R_1 i_1'(t) = E'(t) = -E_o\Omega \operatorname{Sin}\Omega t \\ &L_2 i_2''(t) + M i_1''(t) + R_2 i_2'(t) + 1/C\, i_2(t) = 0\,. \end{aligned} \tag{2}$$

If the circuit is initially dead, then

$$i_1(0) = i_2(0) = 0 \quad \text{and} \quad i_1'(0) = i_2'(0) = 0 \tag{3}$$

and $I_1(s)$, $I_2(s)$, the Laplace transforms of $i_1(t)$, $i_2(t)$, satisfy

$$\begin{aligned} &(L_1s^2 + R_1s)\, I_1(s) + Ms^2 I_2(s) = -E_o\Omega^2/(s^2 + \Omega^2) \\ &Ms^2 I_1(s) + (L_2s^2 + R_2s + 1/C) I_2(s) = 0\,. \end{aligned} \tag{4}$$

We can solve these equations for $I_2(s)$ to obtain

$$I_2(s) = \frac{E_o\Omega^2}{s^2 + \Omega^2}\,\frac{M\,s}{p(s)} \tag{5}$$

where

$$p(s) = (L_1L_2 - M^2)s^3 + (R_1L_2 + R_2L_1)\, s^2 + (R_1R_2 + L_1/C)s + R_1/C\,.$$

Substituting the values of the parameters, we find

$$p(s) = s^3 + 12s^2 + 30s + 100 = (s + 10)(s^2 + 2s + 10)$$

$$= (s + 10)((s + 1)^2 + 3^2).$$

Then, using partial fractions, we can rewrite $I_2(s)$ in the form

$$I_2(s) = E_o \left(\frac{-1}{18} \frac{1}{s + 10} - \frac{1}{170} \frac{11s + 70}{s^2 + 100} + \frac{2}{765} \frac{46s + 37}{s^2 + 2s + 10} \right).$$

Finally, we invert the Laplace transform using Tables 6.1 and 6.2,

$$i_2(t) = E_o \, [2e\text{–}t(46\mathrm{Cos}3t + (37/3)\mathrm{Sin}3t)/765 - e^{-10t}/18$$
$$- (11\mathrm{Cos}10t + 7\mathrm{Sin}10t)/170].$$

Note that the steady state current $i_2(t)$ is equal to

$$i_2(t) = -(E_o/\sqrt{170}) \, \mathrm{Sin}(10t + \varphi),$$

where φ = Arctan(11/7). Then the steady-state voltage drop across the capacitor is given by,

$$1/C \int^t i_2 = (E_o/\sqrt{170}) \, \mathrm{Cos}(10t + \varphi).$$

PROBLEM 7.9

Use Cramer's rule to verify that the component $I_2(s)$ in the solution of the system (4) in the previous problem is given by equation (5) of the previous problem.

SOLUTION 7.9

In order to apply Cramer's rule to find I_2 we must evaluate det A and det A(2) from the system (4), where A(2) denotes the 2 by 2 matrix obtained by replacing the second column of the coefficient matrix A in (4) by the data vector from the right-hand side of the system (4). According to the definition for the determinant of a 2 by 2 matrix, we have

$$\det A = (L_1s^2 + R_1s)(L_2s^2 + R_2s + 1/C) - M^2s^4$$
$$= s[(L_1L_2 - M^2)s^3 + (R_1L_2 + R_2L_1)s^2 + (R_1R_2 + L_1/C)s + R_1/C]$$

and

$$\det A(2) = E_o\Omega^2 \, Ms^2/(s^2 + \Omega^2).$$

Finally, then, Cramer's rule states that

$$I_2(s) = \det A(2)/\det A,$$

which leads to the result found in the previous problem. Note that the difficulty of computing the determinant of an N by N matrix increases very rapidly with increasing N. Thus, Cramer's rule is practical only for small systems.

Linear Algebra

PROBLEM 7.10

Prove (7.12); i.e., show that for any N by N matrix $A = [a_{jk}]$, and all N-vectors X and Y, we have $AX \cdot Y = X \cdot A^* Y$.

SOLUTION 7.10

We use (7.10) and the definition of the inner product to write

$$AX \cdot Y = (a_{11}x_1 + \dots + a_{1N}x_N)\text{¥} + \dots + (a_{N1}x_1 + \dots + a_{NN}x_N)\text{¥}_N$$

$$= (a_{11}\text{¥}_1 + \dots + a_{N1}\text{¥}_N)x_1 + \dots + (a_{1N}\text{¥}_1 + \dots + a_{NN}\text{¥}_N)x_N$$

$$= X \cdot A^* Y.$$

Here A^* denotes the matrix $[\text{å}_{kj}]$, the Hermitian transpose of A.

PROBLEM 7.11

Show that $M^\perp$, the orthogonal complement to a collection M of vectors is a subspace.

SOLUTION 7.11

Let X and Y denote arbitrary vectors in $M^\perp$. Then $X \cdot V=0$ and $Y \cdot V=0$ for all vectors V in the collection M. For arbitrary scalars, α and β,

$$(\alpha X + \beta Y) \cdot V = (aX) \cdot V + (\beta Y) \cdot V$$

$$= \alpha\, X \cdot V + \beta\, Y \cdot V = 0$$

for all vectors V in M; i.e., the linear combination $\alpha X + \beta Y$ belongs to $M^\perp$. This shows that arbitrary linear combinations of vectors in $M^\perp$ must, themselves, belong to $M^\perp$. This proves $M^\perp$ is a subspace.

PROBLEM 7.12

Suppose the vectors $\{u_1, \dots, u_N\}$ form a basis for C^N. Then for each vector X in C^N there exists a unique set of scalars $\{x_1, \dots, x_N\}$ such that

$$X = x_1 u_1 + \dots + x_N u_N . \tag{1}$$

(a) Find the scalars in the case $u_1 = (1,1,1)^T$, $u_2 = (0,1,1)^T$, $u_3 = (0,0,0)^T$ and $X = (1,2,2)^T$.

(b) Show that if the basis is an orthonormal basis, then the scalars are given by $x_k = X \cdot u_k$ for k=1,...,N.

SOLUTION 7.12

(a) In general, for $\{u_1, \dots, u_N\}$ a basis for C^N and X expressed in the form (1), we have

$$(u_1 \cdot u_1)x_1 + \ldots + (u_N \cdot u_1)x_N = X \cdot u_1$$

$$\vdots \qquad (2)$$

$$(u_1 \cdot u_N)x_1 + \ldots + (u_N \cdot u_N)x_N = X \cdot u_N$$

Here we have taken the inner product on both sides of (1) with each of the basis vectors in order. This is a set of N equations in the N unknowns x_1 through x_N. In the special case prescribed in (a) this becomes

$$3x_1 + 2x_2 + x_3 = 5$$

$$2x_1 + 2x_2 + x_3 = 4$$

$$x_1 + x_2 + x_3 = 2 .$$

The solution is easily found to be $x_1 = x_2 = 1$, $x_3 = 0$; i.e., $X = u_1 + u_2$.

(b) If the basis is an orthonormal basis, then it is not necessary to solve a system of equations to find the scalars x_k. In this special case, the coefficient matrix in the system (2) is just the identity matrix and it follows that $x_k = X \cdot u_k$ for k=1,...,N. This result also follows from simply taking the inner product on both sides of (1) with each of the basis vectors in turn. Since the basis vectors are mutually orthogonal, all but one of the inner products on the right is equal to zero.

PROBLEM 7.13

Find the eigenvalues and eigenvectors of the N by N matrix A in (7.16).

SOLUTION 7.13

Writing out the equations in the system $(A - \lambda I)X = 0$, we get

$$-(2+\lambda)x_1 + x_2 = 0$$

$$x_1 - (2+\lambda)x_2 + x_3 = 0$$

$$\ldots$$

$$x_{N-1} - (2+\lambda)x_N = 0$$

This is equivalent to the system of equations (7.17),

$$x_{k-1} - (2+\lambda)x_k + x_{k+1} = 0 \text{ for } k=1,\ldots,N \qquad (1)$$

and

$$x_o = x_{N+1} = 0 . \qquad (2)$$

The assumption $x_k = r^k$ for r not equal to zero reduces (1) to a quadratic equation in the unknown r,

$$r^2 - (2+\lambda)r + 1 = 0. \qquad (3)$$

Equation (3) has roots r_1 and r_2 satisfying

$$r_1 + r_2 = 2 + \lambda \quad \text{and} \quad r_1 r_2 = 1 ; \tag{4}$$

i.e., $(r - r_1)(r - r_2) = r^2 - (r_1 + r_2)r + r_1 r_2$. Then the general solution for x_k is

$$x_k = \alpha r_1{}^k + \beta r_2{}^k \tag{5}$$

for arbitrary constants α and β. The conditions (2) imply that

$$x_o = \alpha + \beta = 0 \tag{6}$$

$$x_{N+1} = \alpha r_1{}^{N+1} + \beta r_2{}^{N+1} = 0. \tag{7}$$

Equation (6) implies $\alpha = -\beta$, and using this in (7) leads to

$$(r_1/r_2)^{N+1} = 1 = e^{i2\pi n}, \text{ n=integer.}$$

But we know from (4) that $r_1 r_2 = 1$ hence it follows that

$$r_1{}^{2N+2} = e^{i2\pi n} \text{ ; so}$$

$$r_1 = e^{i\pi n/(N+1)} \quad \text{and} \quad r_2{}^{-i\pi n/(N+1)} \tag{8}$$

for n an integer. It follows from (4) that the eigenvalues λ of A are given by

$$\lambda_v = r_1 + r_2 - 2 = e^{i\pi n/(N+1)} + e^{-i\pi n/(N+1)} - 2;$$

i.e.,

$$\lambda_n = 2\text{Cos}[n\pi/(N+1)] - 2 \tag{9}$$

Note that for each integer n, λ_n given by (9) is an eigenvalue for A, but (9) yields only N *distinct* values. As n ranges over the values 1 to N, equation (9) produces N distinct values; any other value of n simply repeats one of these N values. But this was predicted by Theorem 7.4. Since A is an N by N matrix, it has N eigenvalues. Moreover, A is a Hermitian matrix since it has all real entries and $a_{jk} = a_{kj}$. It is evident from (9) that the eigenvalues are all real as predicted by Theorem 7.4.

According to (5) and (8), the entries x_k of the eigenvector corresponding to the eigenvalue λ_n are equal to

$$x_k = \alpha[e^{i\pi kn/(N+1)} - e^{-i\pi kn/(N+1)}] = 2i\alpha \text{ Sin}[nk\pi/(N+1)]$$

for k=1,...,N. The parameter α is an arbitrary scalar. Then the eigenvector that corresponds to the eigenvalue λ_n is given by

$$X_{(n)} = C[\text{Sin}(n\pi/(N+1)),\text{Sin}(2n\pi/(N+1)),\ldots,\text{Sin}(Nn\pi/(N+1))]^T \tag{10}$$

for C an arbitrary constant. In obtaining (9) and (10) we have made use of the well-known identities,

$$\text{Sin}\Theta = [e^{i\Theta} - e^{-i\Theta}]/2i \qquad \text{Cos}\Theta = [e^{i\Theta} + e^{-i\Theta}]/2 .$$

PROBLEM 7.14

Find the eigenvalues and eigenvectors for the N by N matrix,

$$A = \begin{bmatrix} -1 & 1 & 0 & \dots & 0 \\ 1 & -2 & 1 & & \\ & & \dots & & \\ & & & -2 & 1 \\ & & & 1 & -1 \end{bmatrix}$$

SOLUTION 7.14

Like the matrix in the previous problem, this is a Hermitian matrix hence A has N real eigenvalues and N corresponding mutually orthogonal eigenvectors. Writing out the equations for the system $(A - \lambda I)X = 0$ produces

$$-(1 + \lambda)x_1 + x_2 = 0$$

$$x_1 - (2 + \lambda)x_2 + x_3 = 0$$

$$\dots$$

$$x_{N-1} - (1 + \lambda)x_N = 0\,.$$

This system of equations in the unknowns $x_1, \dots, x_N$ can be written compactly as

$$x_{k-1} - (2 + \lambda)x_k + x_{k+1} = 0 \quad \text{for } k=1,\dots,N \tag{1}$$

with

$$x_o = x_1 \quad \text{and} \quad x_N = x_{N+1}\,. \tag{2}$$

Then, assuming $x_k = r^k$ as in the previous problem leads to the result that the general solution for x_k is of the form,

$$x_k = \alpha\, r_1{}^k + \beta\, r_2{}^k \quad k=1,\dots,N \tag{3}$$

for arbitrary constants α and β and for r_1 and r_2 satisfying

$$r_1 + r_2 = 2 + \lambda \quad \text{and} \quad r_1 r_2 = 1\,. \tag{4}$$

The condition $x_o = x_1$, together with (3) leads to

$$\alpha(1 - r_1) = \beta(r_2 - 1). \tag{5}$$

Then the second of the conditions (2) implies

$$\alpha(1 - r_1)r_1{}^N = \beta(r_2 - 1)r_2{}^N = \alpha(1 - r_1)r_2{}^N\,.$$

This equation is satisfied if either $r_1 = 1$ or r_1 does not equal 1 and $r_1{}^N = r_2{}^N$. In the case $r_1 = 1$, it follows from (4) that $r_2 = 1$ and $\lambda = 0$. It follows from (3) that the corresponding eigenvector X_0 is given by

$$X_0 = C[1,1,\dots,1]^T, \tag{6}$$

for C an arbitrary constant. For r_1 not equal to 1, we use the fact that $r_1r_2 = 1$ to conclude

$$r_1^{2N} = 1 = e^{i2\pi n} \quad \text{for n=integer.}$$

Then

$$r_1 = e^{in\pi/N} \quad \text{and} \quad r_2 = e^{-1n\pi/N}, \tag{7}$$

and it follows from (4) that

$$\lambda = e^{in\pi/N} + e^{-in\pi/N} - 2 = 2\text{Cos}(n\pi/N) - 2. \tag{8}$$

We must be careful to note that (8) produces N distinct values for λ as the parameter n ranges over the values n=1 to n=N. However, the value $\lambda = -4$, corresponding to n=N is *not an eigenvalue* for A. To see this, we use (5) in (3) to write

$$x_k = \alpha[r_1^k - \frac{1-r_1}{1-r_2} r_1^{-k}]$$

$$= \alpha[r_1^k - \frac{1-r_1}{1-r_2r_1} r_1^{1-k}] = \alpha[r_1^k + r_1^{1-k}],$$

where we have used the fact $r_1 - r_2r_1 = r_1 - 1 = -(1-r_2)$. Then

$$x_k = \alpha r_1^{1/2}[r_1^{k-1/2} + r_1^{1/2-k}]$$
$$= \alpha r_1^{1/2}[e^{i(2k-1)n\pi/2N} + e^{-i(2k-1)n\pi/2N}]$$
$$= 2\alpha r_1^{1/2}\,\text{Cos}[(2k-1)n\pi/2N]. \qquad k=1,\dots,N \tag{9}$$

Note that for n=N, we have $x_k = 0$ for k=1,...,N; i.e., the "eigenvector" X that corresponds to λ_N is the zero vector. Thus λ_N is not an eigenvalue.

The eigenvalues of A are given by (8) for n=0,1,...,N–1. The entries of the corresponding eigenvectors are given by (9) for the same range of n values. That is,

$$X_n = C\,[\text{Cos}(n\pi/2N), \text{Cos}(3n\pi/2N), \dots, \text{Cos}((2N-1)n\pi/2N)]^T \tag{10}$$

for C an arbitrary constant. In particular, note that for n=0, expressions (8) and (10) produce

$$\lambda_o = 0 \quad \text{and} \quad X_o = C[1,1,\dots,1].$$

Systems of Linear Differential Equations

PROBLEM 7.15

Suppose that the N by N matrix A has N independent eigenvectors X_1 to X_N corresponding to eigenvalues λ_1 to λ_N. Show that for a given initial vector G, the solution of the initial value problem

$$d/dt\, U(t) = AU(t), \quad U(0) = \Gamma \tag{1}$$

is given by

$$U(t) = C_1 e^{\lambda_1 t} X_1 + \ldots + C_N e^{\lambda_N t} X_N, \tag{2}$$

where $C = [C_1, \ldots, C_N]^T$ satisfies, $PC = \Gamma$ for P the matrix whose columns are the eigenvectors of A.

SOLUTION 7.15

The assumption that A has N linearly independent eigenvectors implies, via Theorem 7.2, that P^{-1} exists. Then it follows from (7.13) that

$$A = PDP^{-1} \tag{3}$$

and

$$d/dt U(t) = PDP^{-1}U(t).$$

Since $P^{-1}P = I$, and P^{-1} is independent of t, multiplying both sides of this equation by P^{-1} leads to

$$d/dt\, P^{-1}U(t) = DP^{-1}U(t)\,.$$

Now we let

$$V(t) = P^{-1}U(t). \tag{4}$$

Then $V(t)$ solves the initial value problem

$$d/dt V(t) = DV(t), \quad V(0) = P^{-1}\Gamma = C\,. \tag{5}$$

Since the matrix D is a diagonal matrix, each differential equation in the system (5) contains only one unknown function; thus the equations are said to be *uncoupled.* For k=1,...,N

$$d/dt\, v_k(t) = \lambda_k v_k(t), \quad v_k(0) = C\,. \tag{6}$$

Note: the system (1) is a system of *coupled* differential equations; each equation may contain as many as N of the unknown functions. The change of variable (4) reduces the system to the *uncoupled* system (5), which is equivalent to the N single equations (6). Each of the equations (6) is readily solved; the solutions are

$$v_k(t) = C_k e^{\lambda_k t} \tag{7}$$

and

$$V(t) = [C_1e^{\lambda_1 t}, \ldots, C_Ne^{\lambda_N t}]^T \tag{8}$$

It follows from (4) that $U(t) = PV(t)$ and, since the columns of P are the eigenvectors of A, (7.11) implies

$$PV(t) = U(t) = C_1e^{\lambda_1 t}X_1 + \ldots + C_Ne^{\lambda_N t}X_N .$$

This establishes (2) and, incidentally, proves the first part of Theorem 7.5.

PROBLEM 7.16

Suppose the N by N matrix A has N independent eigenvectors X_1 to X_N corresponding to eigenvalues λ_1 to λ_N. Show that for given forcing term $F(t)$, the solution of the initial value problem

$$d/dt\, U(t) = AU(t) + F(t), \quad U(0) = 0 \tag{1}$$

is given by

$$U(t) = \int_0^t e^{\lambda_1(t-\tau)}\, G_1(\tau)\, dt\, X_1 + \ldots \int_0^t e^{\lambda_N(t-\tau)}\, G_N(\tau)\, dt\, X_N \tag{2}$$

where $G(t) = [G_1(t), \ldots, G_N(t)]^T$ is obtained by solving the system $PG = F$. In particular, if the eigenvectors form an orthonormal basis, then $G_k = F \cdot X_k$ for $k=1,\ldots,N$.

SOLUTION 7.16

We shall use variation of parameters to find a particular solution for (1). We assume a particular solution of the form

$$U(t) = \sum_{j=1}^{N} C_j(t)e^{\lambda_j t}\, X_j \tag{3}$$

Since the eigenvectors are independent, they form a basis for C^N, hence we can write

$$F(t) = \sum_{j=1}^{N} G_j(t)X_j \tag{4}$$

It follows from (7.11) that

$$\sum_{j=1}^{N} G_j(t)X_j = PG, \tag{5}$$

where $G = [G_1,\ldots,G_N]^T$, and P is the matrix whose columns are the eigenvectors X_1 to X_N. Then G is obtained by solving $PG = F$.

Now it follows from (3) that

$$d/dt U(t) - AU(t) = \sum_{j=1}^{N} C_j'(t)e^{\lambda_j t}\, X_j ,$$

so the differential equation (1) reduces to

$$\sum_{j=1}^{N} [C_j(t)e^{\lambda_j t} - G_j(t)]X_j = 0. \tag{6}$$

Since the eigenvectors X_j are linearly independent, it follows from (6) that

$$C_j'(t)e^{\lambda_j t} - G_j(t) = 0 \quad \text{for} \quad j=1,...,N$$

and

$$C_j'(t) = \int_0^t e^{-\lambda_j \tau}\, G_j(\tau)\, d\tau \quad \text{for } j=1,...,N \tag{7}$$

Substituting (7) into (3) leads to (2). Since $C_j(0)=0$ for $j=1,...,N$ it follows that the initial condition is satisfied.

Note that if the eigenvectors form an orthonormal basis for C^N, then taking the dot product with X_k on both sides of (4) leads to

$$F(t) \cdot X_k = \sum_{j=1}^{N} G_j(t)\, X_j \cdot X_k = G_k(t).$$

since

$$X_j \cdot X_k = \begin{cases} 0 \text{ if j does not equal k} \\ 1 \text{ if j equals k} \end{cases}$$

Thus, when the eigenvectors of A form an orthonormal basis it is not necessary to solve a system of equations to find the functions $G_k(t)$. This is the case, for example, if A is Hermitian.

PROBLEM 7.17

Solve the initial value problem,

$$d/dt U(t) = \begin{bmatrix} 2 & 3 \\ 1 & 4 \end{bmatrix} U(t), \quad U(0) = [5, 1]^T$$

SOLUTION 7.17

In Problem 7.5 we found that the coefficient matrix for this system had distinct eigenvalues $\lambda_1 = 1$ and $\lambda_2 = 5$, with corresponding independent (but not orthogonal) eigenvectors

$$X_1 = \begin{bmatrix} -3 \\ 1 \end{bmatrix} \quad \text{and} \quad X_2 = \begin{bmatrix} 1 \\ 1 \end{bmatrix}.$$

According to Theorem 7.5, the solution of the initial value problem is

$$U(t) = C_1 e^t X_1 + C_2 e^{5t} X_2,$$

where the constants C_1 and C_2 are obtained by solving, $PC = \Gamma$; i.e.,

$$\begin{bmatrix} -3 & 1 \\ 1 & 1 \end{bmatrix} \begin{bmatrix} C_1 \\ C_2 \end{bmatrix} = \begin{bmatrix} 5 \\ 1 \end{bmatrix}$$

This system may be solved (using Cramer's rule, for example) to obtain $C_1 = -1$ $C_2 = 2$. Then

$$U(t) = 2X_2\, e^{5t} - X_1\, e^t \,;$$

i.e.,

$$\begin{bmatrix} u_1(t) \\ u_2(t) \end{bmatrix} = \begin{bmatrix} 2e^{5t} + 3e^t \\ 2e^{5t} - e^t \end{bmatrix}$$

PROBLEM 7.18

Let B denote the N by N Hermitian matrix that arises in Problem 7.3 in connection with the one-dimensional diffusion model.

(a) Then solve the initial value problem

$$U'(t) = \vartheta B U(t), \quad U(0)=C\,, \tag{1}$$

for C a given initial vector and find the steady state solution.

(b) How is the steady state solution of the initial value problem changed if we replace this matrix by the matrix from Problem 7.4.

(c) Give the physical interpretation of this steady-state behavior in the context of the diffusion models discussed in Problems 7.3 and 7.4.

SOLUTION 7.18

In Problem 7.3 we showed that time dependent one-dimensional diffusion may be modeled by an initial value problem of the form (1). In problem 7.13 we found that the Hermitian matrix B has the following real eigenvalues

$$\lambda_n = 2\text{Cos}[n\pi/(N+1)] - 2 \quad n=1,\ldots,N \tag{2}$$

and corresponding mutually orthogonal eigenvectors

$$X_n = K_n\, [\text{Sin}(n\pi/(N+1)),\text{Sin}(2n\pi/(N+1)),\ \ldots\ ,\text{Sin}(Nn\pi/(N+1))]^T\,. \tag{3}$$

The arbitrary constants K_n can be chosen so that $X_n \cdot X_n = 1$; i.e., so that the eigenvectors are normalized eigenvectors. Then, according to Theorem 7.5, the solution of (1) is given by,

$$U(t) = C_1 \cdot X_1\, e^{\vartheta\lambda_1 t}\, X_1 + \ldots + C_N \cdot X_N\, e^{\vartheta\lambda_N t}\, X_N\,. \tag{4}$$

Note that for each fixed N, the eigenvalues λ_n given by (2) satisfy

$$0 > \lambda_1 > \lambda_2 > \ldots > \lambda_N > -4\,. \tag{5}$$

Then since the parameter ϑ is positive, (5) implies that each of the exponential terms in (4) tends to zero as t tends to infinity. Thus, in the case that B is the matrix from Problem 7.3 the steady state solution of (1) is $U=0$.

(b) When B is the matrix that arises in Problem 7.4, the eigenvalues are as found in Problem 7.14:

$$\lambda_n = 2\mathrm{Cos}(n\pi/N) - 2 \quad n=0,1, \ldots ,N-1 \tag{6}$$

with corresponding eigenvectors

$$X_n = K_n\,[\mathrm{Cos}(n\pi/2N), \mathrm{Cos}(3n\pi/2N), \ldots , \mathrm{Cos}((2N-1)n\pi/2N)]^T . \tag{7}$$

Since B is still Hermitian, the eigenvectors are again mutually orthogonal so if the constants K_n in (7) are chosen such that the eigenvectors are normalized, then the solution to (1) is

$$U(t) = C \cdot X_0 X_0 + C \cdot X_1\, e^{\vartheta\lambda_1 t}\, X_1 + \ldots + C \cdot X_{N-1}\, e^{\vartheta\lambda_{N-1} t}\, X_{N-1} \tag{8}$$

Note that when the eigenvalues are given by (6), then for each fixed value of N, we have

$$0 = \lambda_0 > \lambda_1 > \ldots > \lambda_{N-1} > -4. \tag{9}$$

In particular, $\lambda_0 = 0$ and $\lambda_k < 0$ for k=1,...,N–1. Then as t tends to infinity, the solution $U(t)$ given by (8) tends to the value

$$U = C \cdot X_0 X_0 . \tag{10}$$

Since

$$X_0 = (1/\sqrt{N})\,[1,1, \ldots ,]^T$$

is the normalized eigenvector corresponding to the eigenvalue $\lambda_0 = 0$, it follows that

$$C \cdot X_0 X_0 = [\upsilon,\upsilon, \ldots,\upsilon]$$

where

$$\upsilon = 1/N \sum_{k=1}^{N} C_k = \text{the average of the } C_k\text{'s.}$$

Then the steady state solution (10) may be written

$$U = [\upsilon,\upsilon, \ldots,\upsilon]^T . \tag{11}$$

This is a constant vector in which every component is equal to the average of the initial components.

(c) The matrix B whose eigenvalues were computed in Problem 7.13 is the matrix that arises in Problem 7.3 in modeling diffusion of contaminant in a tube whose ends are in contact with reservoirs in which the concentration of contaminant is equal to zero. In particular, we found the eigenvalues of this matrix are all negative. Then it follows from (5) that no matter what the initial distribution C of contaminant in the cells of the tube, the ultimate (steady-state) distribution will be $U=0$; i.e. the concentration in every cell will tend to zero.

On the other hand, if the ends of the tube are sealed so that no contaminant can escape from the tube (although it is still free to move from cell to cell), then the matrix B in the system (1) is correspondingly changed to the matrix considered in Problem 7.14. In particular, this new matrix has N–1 negative eigenvalues and one eigenvalue equal to zero. As a result, as t tends to infinity, the solution $U(t)$ describing the distribution of contaminant tends to the uniform distribution given by (11). In this state the concentration in every cell in the tube is equal to the value υ, which is the average of the cell concentrations in the initial state C. This is what we would expect in a tube with sealed ends.

It is interesting to note that while the diffusion of contaminant in the tube is modeled by a system of the form (1) for either type of end conditions, the equations describing a tube with open ends lead to a matrix that differs slightly from the matrix that results when the ends of the tube are sealed. The two matrices have slightly different eigenvalues, which accounts in turn for the differences in the steady-state solutions for the corresponding systems.

PROBLEM 7.19

Consider the system of N identical masses considered in Problem 7.5. Solve the initial value problem,

$$X''(t) = \mu BX(t), \tag{1}$$

$$X(0) = C \qquad X'(0) = D \tag{2}$$

where B denotes the N by N matrix considered in Problem 7.3. This corresponds to the situation pictured in Figure 7.4 in which the end masses are attached to rigid supports by springs. The initial conditions (2) specify the initial displacement and initial velocity of each of the masses.

SOLUTION 7.19

The eigenvectors of the Hermitian matrix B are given by (3) in the previous problem. When the constants K_n are chosen such that $X_n \cdot X_n = 1$, these eigenvectors form an orthonormal basis for C^N. If we let P denote the matrix whose columns are these orthonormal eigenvectors, then $A = PDP^{-1}$, where D denotes the diagonal matrix whose diagonal entries are the eigenvalues λ_n of the matrix B. Then the system (1) can be written as

$$X''(t) = \mu PDP^{-1}X(t),$$

or

$$d^2/dt^2(P^{-1}X(t)) = \mu D(P^{-1}X).$$

If we let

$$V(t) = P^{-1}X(t), \tag{4}$$

then the system has been reduced to

$$V''(t) = \mu D V(t)\,. \tag{5}$$

Since the matrix D in the system (5) is a diagonal matrix, each equation in the system contains just a single unknown function; i.e., the system is uncoupled. So, for k=1,...,N

$$d^2/dt^2\, v_k(t) = \mu\lambda_k v_k(t).$$

Recalling from Problem 7.14 that the eigenvalues λ_k of the matrix B are all negative, let

$$\vartheta_k^2 = -\mu\lambda_k \quad \text{for k=1,...,N} \tag{6}$$

Then

$$v_k''(t) = -\vartheta_k^2\, v_k(t) \tag{7}$$

and it follows at once by the techniques of Chapter 4 that

$$v_k(t) = \alpha_k\, \mathrm{Cos}\vartheta_k t + \beta_k\, \mathrm{Sin}\vartheta_k t. \tag{8}$$

Then

$$V(t) = [\alpha_1 \mathrm{Cos}\vartheta_1 t, ..., \alpha_N \mathrm{Cos}\vartheta_N t]^T + [\beta_1 \mathrm{Sin}\vartheta_1 t, ..., \beta_N \mathrm{Sin}\vartheta_N t]^T \tag{9}$$

and $U(t)=PV(t)$. According to (7.11), since the columns of P are the orthonormal eigenvectors of B,

$$\begin{aligned} PV(t) = U(t) &= \alpha_1 \mathrm{Cos}\vartheta_1 t X_1 + ... + \alpha_N \mathrm{Cos}\vartheta_N t X_N + \\ &+ \beta_1 \mathrm{Sin}\vartheta_1 t X_1 + ... + \beta_N \mathrm{Sin}\vartheta_N t X_N\,. \end{aligned} \tag{10}$$

Since the eigenvectors X_k are an orthonormal set, it follows from (2) that

$$\begin{aligned} U(0)\cdot X_k &= \alpha_k = C\cdot X_k \quad \text{for k=1,...,N} \\ U(0)\cdot X_k &= \vartheta_k\, \beta_k = D\cdot X_k \quad \text{for k=1,...,N} \end{aligned} \tag{11}$$

Using these values for α_k and β_k in (10) produces a solution $U(t)$ that satisfies the system of equations (1) and the initial conditions (2).

Note that the system (1) of second-order equations considered in this problem has the same coefficient matrix as the first-order systems considered in the previous problem. While the solutions to the first-order systems of the previous problem decayed steadily with increasing time, the solution to the second-order system (1) oscillates in time with constant amplitude.

PROBLEM 7.20

Solve the inhomogeneous system

$$d/dt \begin{bmatrix} u_1 \\ u_2 \end{bmatrix} = \begin{bmatrix} -2 & 1 \\ 1 & -2 \end{bmatrix} \begin{bmatrix} u_1 \\ u_2 \end{bmatrix} + \begin{bmatrix} 0 \\ t \end{bmatrix}, \quad \begin{bmatrix} u_1 \\ u_2 \end{bmatrix}(0) = \begin{bmatrix} 0 \\ 0 \end{bmatrix}.$$

SOLUTION 7.20

The coefficient matrix is Hermitian with eigenvalues that satisfy

$$(\lambda + 2)^2 - 1 = \lambda^2 + 4\lambda + 3 = (\lambda + 1)(\lambda + 3)$$

Then $\lambda_{1,2} = -1, -3$ and the corresponding orthonormal eigenvectors are

$$X_1 = 1/\sqrt{2}\,[1,1]^T \quad \text{and} \quad X_2 = 1/\sqrt{2}\,[1,-1]^T.$$

According to Theorem 7.5, then $U(t)$ is given by

$$U(t) = \int_0^t e^{-(t-\tau)}\,(F(\tau)\cdot X_1)\,d\tau\, X_1 + \int_0^t e^{-3(t-\tau)}\,(F(\tau)\cdot X_2)\,d\tau\, X_2$$

where

$$F(\tau)\cdot X_1 = -\tau/\sqrt{2} \quad \text{and} \quad F(\tau)\cdot X_2 = \tau/\sqrt{2}.$$

Then

$$U(t) = 1/2 \int_0^t \begin{bmatrix} \tau(e^{-3(t-\tau)} - e^{-(t-\tau)}) \\ \tau(e^{-3(t-\tau)} + e^{-(t-\tau)}) \end{bmatrix} d\tau$$

$$= 1/18 \begin{bmatrix} -6t + 8 + e^{-3t} - 9e^{-t} \\ 12t - 10 + e^{-3t} + 9e^{-t} \end{bmatrix}.$$

This system could be solved by means of the Laplace transform as well.

PROBLEM 7.21

Show that if the matrix A has N linearly independent eigenvectors X_1 to X_N corresponding to eigenvalues λ_1 to λ_N then the solution of (7.20) can be represented as,

$$U(t) = e^{At}\,\Gamma \tag{1}$$

and that the solution of the inhomogeneous (7.22) can be represented as t

$$U(t) = \int_0^t e^{A(t-\tau)}\,F(\tau)\,d\tau \tag{2}$$

SOLUTION 7.21

Since the eigenvectors of A form a basis for C^N, then for any vector Γ we can write

$$\Gamma = C_1X_1 + \ldots + C_NX_N \tag{3}$$

where the scalars C_1 through C_N can be obtained by solving the system

$$PC = \Gamma\,.$$

Here P denotes the matrix whose columns are the eigenvectors of A. It follows from (3) that

$$\begin{aligned} A\Gamma &= C_1AX_1 + \ldots + C_NAX_N \\ &= C_1\lambda_1X_1 + \ldots + C_N\lambda_NX_N \end{aligned}$$

More generally, for any positive integer p,

$$A^p\,\Gamma = C_1\lambda_1{}^pX_1 + \ldots + C_N\lambda_N{}^pX_N\,. \tag{4}$$

Then (4) implies that for any function F(x) that has a convergent power series representation, we have

$$F(A)\Gamma = C_1F(\lambda_1)X_1 + \ldots + C_NF(\lambda_N)X_N \tag{5}$$

In particular, for $F(x)=e^{xt}$, where t denotes a real number, (5) becomes

$$e^{At}\,\Gamma = C_1e^{\lambda_1t}\,X_1 + \ldots + C_Ne^{\lambda_Nt}\,X_N \tag{6}$$

The right-hand side of (6) solves the initial value problem (7.20) and so (1) is verified. Using

$$F(x) = \int_0^t e^{x(t-\tau)}\,d\tau$$

in (5) then leads to (2).

Systems of linear differential equations arise from a number of applications, including electrical networks (as opposed to simple circuits) and mechanical systems with several degrees of freedom. In addition, single equations of higher order can always be reduced, by introducing new dependent variables, to first-order systems. The existence of a unique solution for the basic initial value problem for a first-order system is

guaranteed by Theorem 7.1.

For systems of equations having constant coefficients, the Laplace transform is an effective method of solution provided that the number of equations in the system is small, say only two or three. The Laplace transform reduces the system of differential equations to a system of algebraic equations in the transforms of the unknown functions. When the system is small, Cramer's rule is an efficient way to solve these algebraic equations, particularly if one wants to find only one of the unknown functions rather than all of them. For larger systems, the method rapidly becomes unwieldy.

Constant coefficient systems in many unknown functions are most efficiently solved using the terminology and techniques of linear algebra. In particular, if the eigenvectors of the coefficient matrix form a basis for C^N, *then the solution of the associated system of linear equations can be expressed in terms of the eigenvectors and eigenvalues of A. This representation is possible even when the eigenvectors do not form a basis, but that discussion is not included here. If the coefficient matrix is Hermitian, the eigenvectors not only form a basis, the basis is orthonormal, making the representation even more convenient. This representation often provides significant qualitative information about the solution. For example, if the eigenvalues of the coefficient matrix A are known to be all negative, then the solution of the first-order system* $U'(t)=AU$ *must tend to* 0 *as t tends to infinity.*

In the case of certain Hermitian matrices that occur frequently in a variety of applications, it is possible to compute the eigenvalues and eigenvectors by reducing the eigenvalue problem to a simple difference equation. Problems 7.13 and 7.14 illustrate this reduction procedure, and Problem 7.18 provides an example of the application of the results.

SUPPLEMENTARY PROBLEMS

Solve the systems by means of the Laplace transform

7.1 $(D + 2)x_1(t) + x_2(t) = 1 \qquad x_1(0)=1,$

$(D - 1)x_1(t) + \quad + Dx_3(t) = -1 \qquad x_2(0)=0,$

$(D - 1)x_2(t) + (D + 2)x_3(t) = 0 \qquad x_3(0)=0\,.$

7.2 $(D^2 - 1)x_1(t) - D^2x_2(t) = -2\text{Sint}$, $x_1(0)=A_1$ $x_1'(0)=B_1$

$(D^2 + 1)x_1(t) - D^2x_2(t) = 0$, $x_2(0)=A_2$ $x_2'(0)=B_2$.

7.3 $(D^2 + 1)x_1(t) + x_2(t) = 0$, $x_1(0)=x_1'(0)=0$

$Dx_1(t) + Dx_2(t) = 0$ $x_2(0)=1$

$$d/dt \begin{bmatrix} x(t) \\ y(t) \end{bmatrix} = \begin{bmatrix} 1 & 2 \\ 3 & 2 \end{bmatrix} \begin{bmatrix} x(t) \\ y(t) \end{bmatrix}$$

7.4 $D^2x(t) + Dy(t) = \text{Cost}$ $x(0)=-1$ $x'(0)=-1$

$$d/dt \begin{bmatrix} x(t) \\ y(t) \\ z(t) \end{bmatrix} = \begin{bmatrix} 1 & 2 & 1 \\ 6 & -1 & 0 \\ -1 & -2 & -1 \end{bmatrix} \begin{bmatrix} x(t) \\ y(t) \\ z(t) \end{bmatrix}$$

$-x(t) + D^2y(t) = \text{Sint}$ $y(0)= 1$ $y'(0)=0$

$$A = \begin{bmatrix} -1 & 1 & 0 & \cdots & 0 \\ 1 & -2 & 1 & & \\ & & \cdots & & \\ & & & -2 & 1 \\ & & & 1 & -2 \end{bmatrix}$$

7.5 $D^2x(t) = y(t) + 2$, $D^2y(t) = x(t) - 2$

$$d/dt \begin{bmatrix} x(t) \\ y(t) \end{bmatrix} = \begin{bmatrix} 0 & 1 \\ 8 & -2 \end{bmatrix} \begin{bmatrix} x(t) \\ y(t) \end{bmatrix} + \begin{bmatrix} 0 \\ e^t \end{bmatrix} \quad \begin{bmatrix} x(0) \\ y(0) \end{bmatrix} = \begin{bmatrix} 1 \\ -4 \end{bmatrix}$$

7.6 Find a general solution for the system of equations,

7.7 Find the general solution for,

7.8 Find the eigenvalues and eigenvectors for the N by N matrix,

7.9 Solve the system of equations

ANSWERS TO SUPPLEMENTARY PROBLEMS

7.6 $\begin{bmatrix} x(t) \\ y(t) \end{bmatrix} = C_1 \begin{bmatrix} 1 \\ -1 \end{bmatrix} e^{-t} + C_2 \begin{bmatrix} 2 \\ 3 \end{bmatrix} e^{4t}$

7.1 $x_1(t) = 1-e^{-t}+e^{-2t}$, $x_2(t)= -1+e^{-t}$, $x_3(t) = (-1+4e^{-t}-3e^{-2t})/2$

7.7 $\begin{bmatrix} x(t) \\ y(t) \\ z(t) \end{bmatrix} = \begin{bmatrix} 1 \\ 6 \\ -13 \end{bmatrix} + C_2 \begin{bmatrix} 1 \\ -2 \\ -1 \end{bmatrix} e^{-4t} + C_3 \begin{bmatrix} 2 \\ 3 \\ -2 \end{bmatrix} e^{3t}$

7.2 $x_1(t) = \text{Sin}t + (B_1-1)t + A_1$, $x_2(t) = A_2+B_2t+(1-B_1)t^3/6-A_1t^2/2$.

7.8 $\lambda_n = 2[\text{Cos}(\frac{2n-1}{2N+1}\pi) - 1]$ $n=1,2,\ldots,N$

$X_n = [\text{Cos}(\frac{2n-1}{2}\pi x_1), \text{Cos}(\frac{2n-1}{2}\pi x_2), \text{Cos}(\frac{2n-1}{2}\pi x_N)]^T$

7.9 $\begin{bmatrix} x(t) \\ y(t) \end{bmatrix} = \begin{bmatrix} 31 \\ 124 \end{bmatrix} \frac{e^{-4t}}{30} + \begin{bmatrix} 1 \\ 2 \end{bmatrix} \frac{e^{2t}}{6} + \begin{bmatrix} -1 \\ -1 \end{bmatrix} \frac{e^{t}}{5}$

7.4 $x(t) = -\text{Cos}t - \text{Sin}t$ $y(t) = \text{Cos}t$

7.5 $x(t) = A\text{Sin}t + B\text{Cos}t + Ce^{t} + De^{-t} + 2$

$y(t) = -A\text{Sin}t - B\text{Cos}t + Ce^{t} + De^{-t} - 2$

8

Systems of Nonlinear Equations

In the previous chapter we confined our attention to systems of linear differential equations. We showed how such systems arise naturally in many applications, and we presented methods of solving linear systems having constant coefficients.

In this chapter we shall see applications that lead to systems of nonlinear differential equations. There are many such applications, and in fact physical systems that are modeled by nonlinear problems probably outnumber systems that lead to linear models. Unfortunately there are no general methods for solving nonlinear systems of differential equations and so introductory treatments of the subject of differential equations tend to limit their discussion of nonlinear systems.

We shall see that systems of differential equations fall into two classes: autonomous and nonautonomous. Since the behavior of nonautonomous systems can be more complicated than that of autonomous systems, we shall concentrate most of our attention on autonomous systems. Even when we are unable to solve autonomous systems of nonlinear equations, we can discover considerable information of a qualitative nature about the solution. In particular, we can determine how the solution behaves in neighborhoods of special points called critical points, and by piecing together these local pictures, we can often obtain a good global picture of how a given system acts. The local analysis of the nonlinear problem is carried out by studying an associated linear problem so the methods of the previous chapter come into play. However, we shall see that there are a number of important ways in which nonlinear problems differ from problems that are linear. For exam-

ple, only nonlinear systems exhibit so called limit cycle behavior and only a system that is nonlinear can behave chaotically.

DYNAMICAL SYSTEMS

We consider systems of differential equations of the form

$$dx_1(t)/dt = P_1(x_1,\ldots,x_N)$$

$$\vdots \qquad (8.1)$$

$$dx_N(t)/dt = P_N(x_1,\ldots,x_N)$$

where for k=1,...,N, P_K denotes a given smooth function of N variables, and the $x_K(t)$ are the unknown solutions being sought. The system (7.20) in the previous chapter is a special case of (8.1) in which each of the functions P_K is linear.

Autonomous and Non-autonomous Systems

We shall devote the majority of our attention to systems (8.1) in the case that N = 2. Much of the interesting behavior that is present in systems with N > 2 is already present in two-dimensional sytems (i.e., systems with N = 2), and when N = 2 we can visualize the solution behavior by plotting solution curves in the x_1x_2 plane. Since the number of equations and unknowns is only two, it will be more convenient to dispense with the subscript notation and consider systems of the form

$$\begin{aligned} x'(t) &= P(x,y) \\ y'(t) &= Q(x,y) \end{aligned} \qquad (8.2)$$

We refer to the system (8.2) as an *autonomous* system, as opposed to a system of the form

$$\begin{aligned} x'(t) &= P(x,y,t) \\ y'(t) &= Q(x,y,t) \end{aligned} \qquad (8.3)$$

which is said to be *nonautonomous*. The functions P and Q in (8.3) depend *explicitly* on the independent variable t, while the functions in (8.2) depend on t only implicitly through the dependent variables x and y. Autonomous systems are usually more easily analyzed than nonautonomous systems, although many of the results that are true for the autonomous case can be generalized to the nonautonomous case as well.

Example 8.1 Autonomous and Non-autonomous Systems

8.1(A)

Let K,τ denote given constants. The system

$$x'(t) = y \qquad x(\tau) = K$$

$$y'(t) = -x \qquad y(\tau) = 0$$

is autonomous; i.e., the functions $P(x,y) = y$ and $Q(x,y) = -x$ do not explicitly involve the independent variable t. In an autonomous system it is always possible to eliminate the independent variable t by

$$dx/P(x,y) = dt = dy/Q(x,y).$$

In this example, that leads to

$$dx/y = -dy/x \quad \text{or} \quad xdx + ydy = 0.$$

We can integrate this last expression to obtain the relationship

$$x^2(t) + y^2(t) = \text{constant}.$$

This expression is a first integral for the system, but it is not a solution since it does not explicitly provide x and y as functions of t. We shall see, however, that a first integral provides useful information about the behavior of the system.

8.1(B)

The following system is nonautonomous:

$$x'(t) = x/t \qquad x(\tau) = K$$

$$y'(t) = 2t \qquad y(\tau) = 0,$$

since $P(x,y,t) = x/t$ and $Q(x,y,t) = 2t$ depend explicitly on t. In general we cannot eliminate the independent variable in a nonautonomous system, but in this example we can integrate each of the equations independently to obtain the solution,

$$x(t) = K\tau/t \text{ and } y(t) = t^2 - \tau^2.$$

Orbits

For given smooth functions $x = x(t)$ and $y = y(t)$, and real parameter τ, the set of points

$$\Gamma = \{(x,y) : x = x(t),\ y = y(t),\ t > \tau\}$$

describes a curve in the xy plane. The curve Γ is said to be an *orbit* of the system (8.2) if the functions $x = x(t)$, $y = y(t)$ are solutions for the equations in (8.2) for $t > \tau$. If $x(\tau) = x_0$ and $y(\tau) = y_0$ then we say that Γ is the orbit through the point (x_0, y_0).

We refer to systems of differential equations like (8.1), (8.2) and (8.3) as *dynamical systems* and refer to the xy plane as the *phase plane* when plotting orbits of 2-dimensional dynamical systems. Note that distinct orbits of autonomous systems never intersect; i.e., at each point of an orbit Γ, there is a unique tangent direction defined by the autonomous system. If two orbits were to intersect, there would be *two* tangent directions associated with a single point. Similarly, an orbit of an autonomous dynamical system does not cross itself unless that orbit is a simple closed curve. In that case the closed orbit represents a periodic solution to the system of equations. Examples illustrating the various types of orbits are provided in the solved problems.

Example 8.2 Orbits of Autonomous and Non-autonomous Systems

8.2(A)

In the autonomous system considered in Example 8.1, we eliminated the independent variable t in order to obtain the relation

$$x(t)^2 + y(t)^2 = \text{constant for all } t > \tau .$$

Then the initial conditions imply that the circle $x^2 + y^2 = K^2$ is an orbit for this autonomous system. In particular, this circle is the orbit through the point (K,0). As the parameter K varies, the orbits form a family of concentric circles of radius K. Note that each point in the phase plane is on one and only one circle. The fact that the orbits are closed curves indicates that the solution of the system is periodic. Since this example is linear, the methods of the previous chapter can be used to show that the periodic solution is $x(t) = K\text{Cos}(t - \tau)$, $y(t) = -K\ \text{Sin}(t-\tau)$. In more difficult examples we may not be able to construct the solution, but if we can show that the orbits are closed curves, then we know that the system has periodic solutions.

8.2(B)

The independent variable cannot be eliminated from the nonautonomous system considered in Example 8.1. However, direct integration of the two equations lead to the solution

$$x(t) = Kt/\tau \text{ and } y(t) = t^2 - \tau^2 .$$

We can eliminate t from the solution, by

$$y = t^2\,(x^2 - K^2)/K^2.$$

Thus, for a fixed value of τ, the orbits for the nonautonomous system form a family of parabolas as the parameter K is varied. However, the shape of the parabolic orbit through (K,0) varies with the value of the initial time τ at which the orbit is required to pass through the prescribed point (K,0). Note also that for a fixed value of τ, the orbits through (K,0) for various K all meet at the point $(0, -\tau^2)$.

This example illustrates that even in very simple cases, the behavior of nonautonomous systems is more complicated than the behavior of autonomous systems. For this reason, and since autonomous systems arise frequently in applications, we shall consider only autonomous systems in what follows.

LINEARIZATION

Critical Points

Although it is generally not possible to solve nonlinear systems, we can often learn a great deal about the solution by examining the orbits near special points called *critical points*.

Definition 8.1 The point (x_c,y_c) is a *critical point* for the autonomous system (8.2) if

$$P(x_c, y_c) = Q(x_c, y_c) = 0.$$

Note that at a critical point $x'(t) = y'(t) = 0$, so a point located at a critical point is at rest. Additionally, we can show that a point moving along an orbit that passes through a critical point cannot reach the critical point in finite time. Note also that for a *linear* system the one and only critical point is the origin, $x = y = 0$.

APPROXIMATION BY A LINEAR SYSTEM

Suppose that (x_C,y_C) is a critical point for the system (8.2). We can approximate P(x,y) and Q(x,y) in a neighborhood of the critical point by the first few terms of the two-dimensional Taylor's series; i.e.

$$\begin{aligned} P(x,y) &\sim P(x_c,y_c) + P_x(x_c,y_c)(x-x_c) + P_y(x_c,y_c)(y-y_c) \\ Q(x,y) &\sim Q(x_c,y_c) + Q_x(x_c,y_c)(x-x_c) + Q_y(x_c,y_c)(y-y_c)\,. \end{aligned} \tag{8.4}$$

Here P_x, P_y, Q_x, and Q_y denote the partial derivatives of P and Q with respect to x and y. Then, since x_c and y_c are constants, and $P(x_c,y_c) = Q(x_c,y_c) = 0$, we can approximate the system (8.2) in a neighborhood of the critical point (x_c,y_c) by the linear system

$$\begin{aligned} d/dt(x(t)-x_c) &\sim P_x(x_c,y_c)(x-x_c) + P_y\,(x_c,y_c)(y-y_c) \\ d/dt(y(t)-y_c) &\sim Q_x(x_c,y_c)(x-x_c) + Q_y\,(x_c,y_c)(y-y_c) \end{aligned}$$

JACOBIAN MATRIX

In matrix notation this becomes

$$dZ/dt = J(x_c,y_c)Z(t) \qquad (8.5)$$

where $Z(t) = [x(t)-x_c, y(t)-y_c]^T$ and $J(x_c,y_c)$ denotes the *Jacobian* matrix

$$J(x_c,y_c) = \begin{bmatrix} P_x(x_c,y_c) & P_y(x_c,y_c) \\ Q_x(x_c,y_c) & Q_y(x_c,y_c) \end{bmatrix} \qquad (8.6)$$

The linear system (8.5) is called the *linearization* of (8.2) about the critical point (x_c,y_c). Since $Z = 0$ corresponds to $x = x_c$ and $y = y_c$, it follows that the orbits of the system (8.5) near $Z=0$ may approximate the orbits of (8.2) near (x_c,y_c). As we shall now see, there are a limited number of possible configurations for the orbits of (8.5) near $Z = 0$.

Orbits of Linear Systems

Consider the linear system

$$X'(t) = AX(t) \qquad (8.7)$$

where A denotes a 2 by 2 matrix with real, constant entries. We suppose that A is nonsingular with eigenvalues λ_1 and λ_2 and corresponding eigenvectors $X_1$and X_2. Since A is nonsingular, neither of the eigenvalues is zero. If the eigenvectors are independent, then the general solution of (8.7) is given by

$$\text{(a)} \qquad X(t) = C_1 e^{\lambda_1 t}X_1 + C_2 e^{\lambda_2 t}X_2, \qquad (8.8)$$

or, if the eigenvalues form a complex conjugate pair $\lambda_{1,2} = \upsilon \pm i\eta$ the solution may be written in the form

$$\text{(b)} \qquad X(t) = e^{\upsilon\tau}(C_1 \mathrm{Cos}\eta t + C_2 \mathrm{Sin}\eta t)$$

The origin is the only critical point for (8.7), and the behavior of the orbits is determined by the character of the eigenvalues λ_1 and λ_2. There are only the following cases to consider:

REAL EIGENVALUES

UNSTABLE NODE $\quad 0 < \lambda_1 < \lambda_2$

In this case, the positive eigenvalues cause the length of $X(t)$ to increase without bound as t increases; i.e. all orbits move away from the origin. We say that the origin is an *unstable node* when both eigenvalues are real and positive.

SADDLE POINT $\quad \lambda_1 < 0 < \lambda_2$

Since λ_1 is negative, orbits approach the origin in the direction of the eigenvector X_1. Similarly, orbits escape to infinity in the direction X_2 corresponding to the positive eigenvalue. We say the origin is a *saddle point* when the eigenvalues are real and opposite in sign.

STABLE NODE $\lambda_1 < \lambda_2 < 0$

When the eigenvalues are real and both negative, the length of $X(t)$ decreases to zero with increasing t; i.e., all orbits approach the origin. We say that the origin is a stable node.

DEGENERATE NODE $\lambda_1 = \lambda_2 = \lambda$ with $\dim NS[A - \lambda I] = 1$

If the eigenvalue is repeated and there is only a single eigenvector X, the orbits approach the origin or escape along the direction X, according to whether λ is negative or positive. We can think of this as the *degenerate* case that occurs when the eigenvectors X_1 and X_2 "scissor" together to form a single vector in the direction X.

Typical orbit configurations for each of these four cases are pictured in the figures in the solved problems.

When the entries of the matrix A are real, then *complex* eigenvalues can occur only as complex conjugate pairs.

COMPLEX EIGENVALUES $\lambda_{1,2} = \upsilon \pm i\eta$

CENTER $\upsilon = 0$

If $\upsilon=0$, then the solution is periodic; i.e. the orbits are closed curves. We say that the origin is a *center* in this case.

STABLE FOCUS $\upsilon < 0$

When υ is negative, the periodic components of the solution are multiplied by a decreasing exponential factor, $e^{\upsilon t}$. Then the orbits spiral in toward the origin. We say that the critical point is a *stable focus*.

UNSTABLE FOCUS $\upsilon > 0$

For υ positive the amplitude of the periodic oscillation grows exponentially and the orbits spiral outward, away from the origin. We say that we have an *unstable focus*.

Typical orbits for each of these cases are pictured in figures in the solved problems.

STABILITY

Equilibrium States

Many physical systems are modeled by autonomous systems of differential equations. For such systems a critical point represents an *equilibrium* state of the physical system. Once the system has been placed in a state corresponding to a critical point, it will remain in that state. Some equilibrium states are such that if the system is disturbed slightly when it is in the equilibrium state, the system returns to the equilibrium state. Such states are said to be *stable*. If a slight disturbance of the system in an equilibrium state causes the system to assume a different state, then the state is said to be *unstable*. A motionless pendulum hanging straight down is in a stable equilibrium state. If the pendulum is pulled slightly to one side it swings back and forth until friction eventually brings it to rest again hanging motionless straight down. A long rod, balanced on end is in an unstable equilibrium state. Pushing the rod to one side will cause it to topple and come to rest in a new equilibrium state. For mathematical purposes, we need more precise definitions of these terms.

ASYMPTOTIC STABILITY

Let $\Gamma = \{(x,y):x = x(t),\ y = y(t)\}$ denote an orbit for the system (8.2) and suppose that (x_c,y_c) is a critical point for this system. For $(x(t),y(t))$ on Γ, let

$$d(t) = \sqrt{(x(t)-x_c)^2 + (y(t)-y_c)^2}\ ;$$

i.e., $d(t)$ is the distance between the critical point and the point $(x(t),y(t))$ on the orbit Γ.

Definition 8.2 The critical point (x_c,y_c) is said to be:

(a) *stable* if for every $\varepsilon > 0$ there exists a $\delta > 0$ such that

$$d(0) < \delta \text{ implies } d(t) < \varepsilon \text{ for all } t > 0.$$

(b) *asymptotically stable* if it is stable and, in addition, there is an $\eta>0$ such that

$$d(0) < \eta \text{ implies that } d(t) \to 0 \text{ as } t \to \infty$$

If a critical point is stable, then any orbit that begins close to the critical point will remain near the critical point. The critical point is asymptotically stable if it is stable and, in addition, any orbit that starts out close to the critical point is eventually "attracted to" the critical point. An asymptotically stable critical point is said to be an *attractor* for all orbits sufficiently near it. Any critical point that is not stable is said to be *unstable*.

A stable node and stable focus are asymptotically stable critical points for a linear autonomous system. A center is stable but not asymptotically stable. The unstable node and focus and the saddle are examples of unstable critical points for linear autonomous systems. As to the stability of critical points for nonlinear autonomous systems, we have the following theorem.

Theorem 8.1

Theorem 8.1 Let (x_c,y_c) be a critical point for the autonomous system (8.2) and let (8.5) denote the linearization of (8.2) about this critical point. If $Z = 0$ is an asymptotically stable critical point for (8.5), then (x_c,y_c) is an asymptotically stable critical point for (8.2). If $Z = 0$ is an unstable critic point for (8.5), then (x_c,y_c) is an unstable critical point for (8.2).

When $Z = 0$ is a stable but not asymptotically stable critical point for (8.5), the theorem says nothing about the stability or instability of (x_c, y_c) for (8.2). We say then that (x_c, y_c) is *neutrally stable.* More delicate tests of stability are required in neutrally stable cases.

LIMIT CYCLES

There are a limited number of possibilities for the behavior of orbits of *linear* autonomous systems:

1. the orbit is attracted to the origin;
2. the orbit is repelled by the origin;
3. the orbit is a closed curve around the origin.

For *nonlinear* autonomous systems there is an additional possibility:

4. the orbit is attracted to a closed curve around the critical point.

In this case the closed curve attracting the orbit is called a *limit cycle.* Limit cycles do not occur for linear systems. Existence of limit cycles is suggested by the following theorem.

Theorem 8.2 Poincare-Bendixson Theorem

Theorem 8.2 Let U denote a closed, bounded region of the phase plane containing no critical points for the system (8.2). If Γ is an orbit of the system (8.2) which originates in U and remains in U for all t, then either Γ is a closed curve or else Γ spirals toward a closed curve in U. In this latter case, the closed curve attracting Γ is called a *limit cycle.*

STRANGE ATTRACTORS

The bounded region referred to in this famous theorem is an annular region containing the limit cycle. The critical point lies in the disc surrounded by the annulus. The limit cycle and asymptotically stable critical points are two examples of *attractors;* point sets to which orbits of dynamical systems are attracted.

The possible modes of behavior for the orbits of an autonomous system in the plane are limited to the ones listed here. In particular, points and closed curves are the only possible attractors in the plane. For autonomous systems in more than two dimensions, more complicated attractors may occur. These attracting point sets may be neither points nor curves but, in fact, complicated sets whose dimension may not even be an integer. These sets are called *strange attractors.* The orbits of a nonlinear autonomous system having a strange attractor may appear to be very erratic since they are approaching this geometrically unusual set. The word *chaotic* is often used to describe the behavior of such systems. Analysis of chaotic dynamical systems is beyond the scope of this text.

In the solved problems we plot orbits for many autonomous nonlinear systems, including several nonlinear systems where we are not able to construct the solutions. This has been accomplished by constructing *numerical solutions* for the systems using techniques that will be developed in Chapter ten.

SOLVED PROBLEMS

Orbits of Linear Systems

PROBLEM 8.1

Equation (5.1) states that x(t), the unforced displacement of a mass on a damped spring, satisfies

$$mx''(t) = -Kx(t) - Cx'(t). \tag{1}$$

Letting $y(t) = x'(t)$ reduces this second-order equation to a first order system:

$$x'(t) = y(t)$$
$$y'(t) = -K/m\, x(t) - C/m\, y(t).$$

(a) For $C = 0$, the undamped case, plot the orbits through $(a_k,0)$, $k = 1,2,3$ where $0<a_1<a_2<a_3$.

(b) For $0<C^2<4Km$, the underdamped case, plot the orbits through the points $(0,a_k)$ $k = 1,2,3$.

(c) Show that the critically damped case, $C^2 = 4Km$, corresponds to a stable, degenerate node at the origin. Plot several orbits.

(d) Show that the overdamped case, $C^2>4Km$, corresponds to a stable node at the origin. Plot several orbits.

SOLUTION 8.1

The system (2) is a linear system with coefficient matrix

$$A = \begin{bmatrix} 0 & 1 \\ -K/m & -C/m \end{bmatrix}.$$

The origin is the single critical point for a linear system and the character of the critical point is determined by the eigenvalues of A. If we let $C/m=2b$ and $K/m=\Omega^2$ then the eigenvalues of A are equal to

$$\lambda_{1,2} = -b \pm i\sqrt{\Omega^2 - b^2} = -b \pm i\omega_D. \tag{3}$$

8.1(a) If C=0, then b=0 and the eigenvalues in this case are $\lambda_{1,2} = \pm\, i\Omega$. Since the eigenvalues of A are complex with real part equal to zero, the critical point at the origin is a *center*, and the orbits are going to be closed curves. To see what the orbits are, note that for b=0, the system (2) has the form

$$x'(t) = y$$
$$y'(t) = -\Omega^2 x\,.$$

Multiplying the first equation by $2\Omega^2 x$ and the second by 2y, we can add them to obtain

$$2\Omega^2 xx' + 2\,yy' = d/dt\,(\Omega^2 x^2 + y^2) = 0.$$

It follows from this that

$$\Omega^2 x^2 + y^2 = \text{Constant.} \tag{4}$$

The expression (4) is a *first integral* for the equation (1) and the system (2). For each choice of a value for the constant, the locus of points (x,y) in the plane for which (4) is satisfied is an ellipse. Each such ellipse is an orbit for the system (2). Choosing the constant in (4) such that

$$\text{Constant} = \Omega^2 a_k + 0 \text{ for } k = 1,2,3$$

produces the elliptic orbit through the points $(a_k, 0)$ k = 1,2,3. These initial states correspond to the situation in which the mass is given an initial displacement equal to a and is then released from rest. The orbits describe the subsequent motion of the spring-mass system. These orbits are pictured in Figure 8.1.

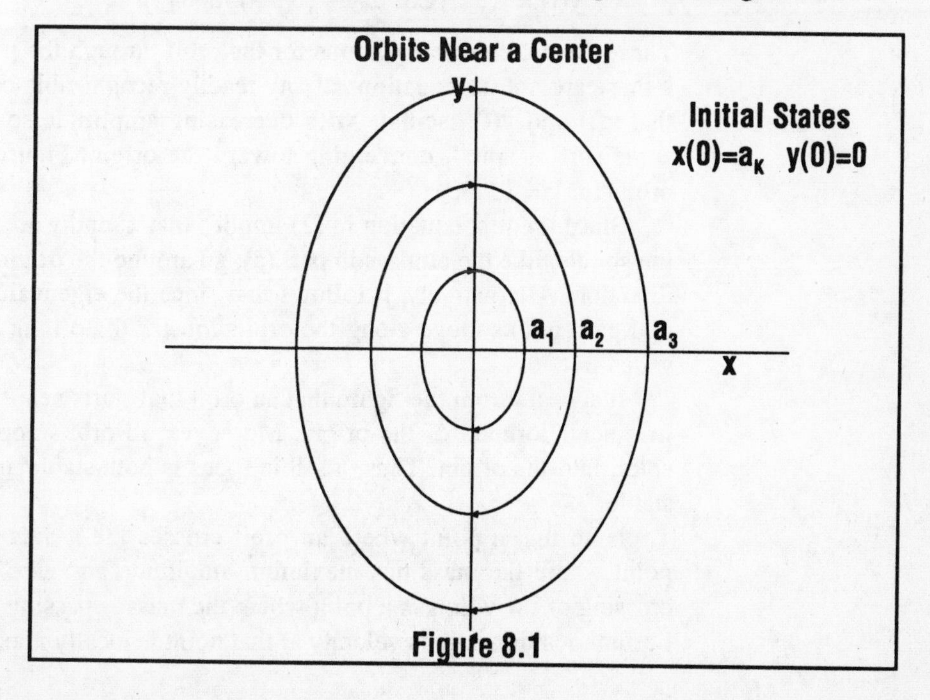

Figure 8.1

Note that there are two points on each of the three orbits where $x=0$. According to the equation for the orbit,

$$\Omega^2x^2+y^2 = \Omega^2a_k^2 \qquad k = 1,2,3\,, \tag{5}$$

these occur when $y = \pm\ \Omega a_k$, $k = 1,2,3$. These points on the orbit correspond to the instant when the mass passes through the equilibrium position, moving first in one direction and then in the other. Note that the derivative x' has the same sign as y. Thus x is increasing when y is positive and decreasing when y is negative. It follows that the elliptical orbits must be traversed in the *clockwise* direction.

Note also that there are two points on each orbit where $y = 0$. It follows from (4) that these occur when $x = \pm\ a_k$ i.e., the velocity of the oscillating mass equals zero when the mass is at the points of extreme amplitude on either side of the equilibrium position.

It is evident from the figure that orbits that start near the origin, remain in a neighborhood of the origin. However, an orbit that starts at a point outside the origin remains a positive distance from the origin for all time. Thus, a center is stable but it is not aysmptotically stable.

8.1(b) If $0 < C^2 < 4Km$ then $\Omega^2 > b^2$ and, according to (3), the eigenvalues of A are complex with negative real part. Then the critical point at the origin is a *stable focus.* We can solve this linear system with the initial conditions $x(0) = 0$, $y(0) = a_k$, to obtain

$$\begin{aligned} x(t) &= (a_k/\omega_D)e^{-bt}\,\text{Sin}\ \omega_D t. \\ y(t) &= a_k e^{-bt}\,[\text{Cos}\ \omega_D t - (b/\omega_D)\text{Sin}\ \omega_D t]. \end{aligned} \tag{6}$$

These are parametric equations for the orbit through the point $(0,a_k)$. While k these are not the equations of any readily recognizable curve, it is evident that x(t) and y(t) oscillate with decreasing amplitude so that the curve is some sort of spiral, converging toward the origin. Figure 8.2 pictures the orbits for $k = 1,2,3$.

Since the first equation in (2) implies that x' and y are of the same sign, the spirals, like the ellipses in part (a), go around the origin in the clockwise direction. Alternatively, it follows that since the eigenvalues have *negative* real part, points move along the orbits *toward* the origin; i.e., in the clockwise direction.

It appears from the figure that an orbit that starts near the origin remains in a neighborhood of the origin. Moreover, all orbits appear to eventually spiral into the origin. Thus, a stable focus is both stable and asymptotically stable.

Note that a point where an orbit crosses the x-axis corresponds to a point where the mass has maximum amplitude and zero velocity, while a crossing of the y-axis is a point where the mass is passing through the equilibrium position and its velocity at that point is locally maximal.

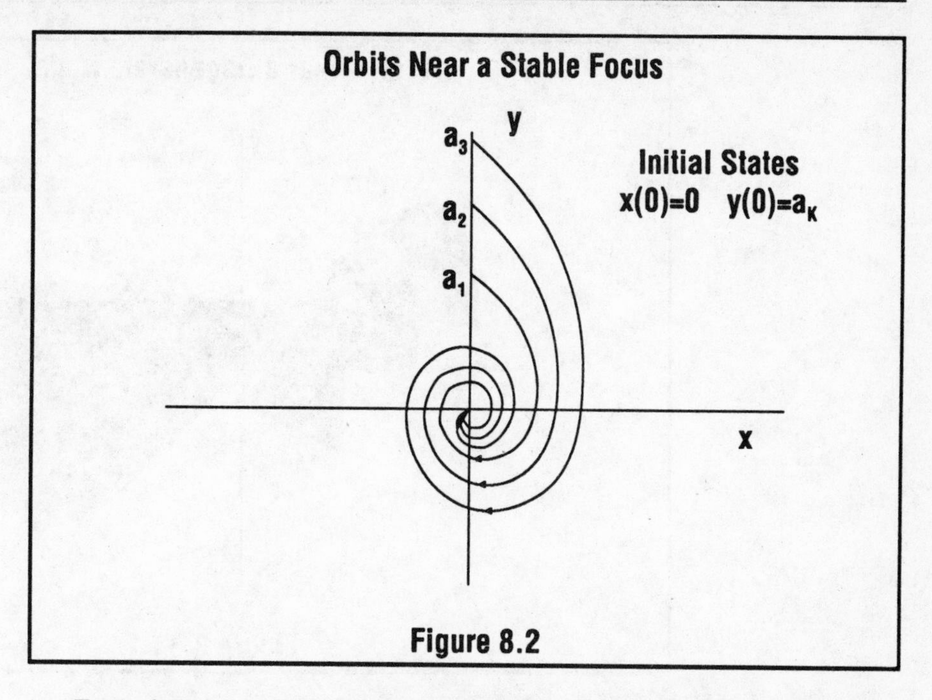

Figure 8.2

Each time one of the spiral orbits in figure 8.2 crosses the x-axis, the crossing is closer to the origin than the previous crossing. This corresponds to the fact that the damped spring-mass system oscillates with decreasing amplitude. Similarly, each time one of the spiral orbits crosses the y-axis, the crossing is closer to the origin than the previous crossing of the y-axis; i.e., the oscillating mass is slowing down due to damping.

8.1(c) If $C^2 = 4Km$ then $\Omega^2 = b^2$ and it follows from (3) that the only eigenvalue for A is the negative number $\lambda = -b$. It is not difficult to show that dimNS[A + bI] is equal to one, hence the origin is a *degenerate, stable node* for the system (2). We can easily solve the linear system in this case to obtain

$$x(t) = C_1 e^{-\Omega t} + C_2\, t e^{-\Omega t}$$

$$y(t) = -\,\Omega C_1\, e^{-\Omega t} + C_2\, e^{-\Omega t}(1-\Omega t).$$

Once again, for each choice of C_1 and C_2 these are the parametric equations of an orbit for the system (2), although the equations are not readily recognizable as the equations of any simple curve. Several orbits are plotted in Figure 8.3, and since λ is negative, the direction of motion along any orbit is always toward the origin. It may help to refer to solved Problems 5.1 and 5.2 at this point in order to reconcile the phase plane portrait in Figure 8.3 with the motion of a critically damped system.

8.1(d) In the overdamped case, $C^2 > 4Km$, we have $b^2 > \Omega^2$ and then it follows from (3) that A has two real, negative eigenvalues

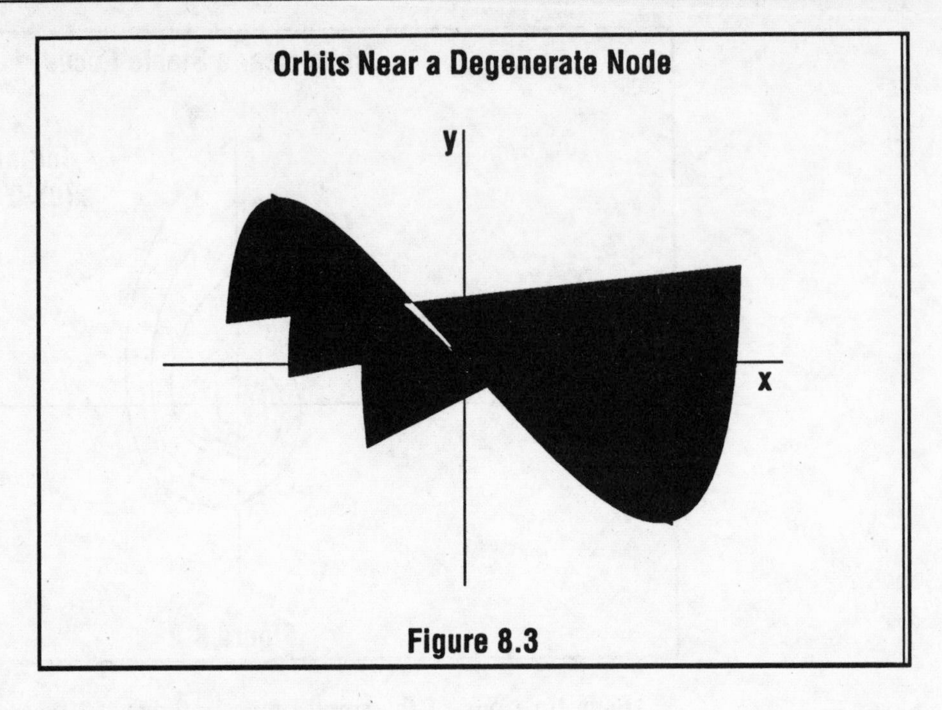

Figure 8.3

$$\lambda_1 = -b + \sqrt{b^2 - \Omega^2} < 0$$

$$\lambda_2 = -b - \sqrt{b^2 - \Omega^2} < \lambda_1.$$

The solution in this case is given by equation (8.8)(a) and, since both eigenvalues are negative, it follows that $X(t)$ tends to $0 = (0,0)$ as t tends to infinity; i.e., all orbits approach the origin, which is a *stable node*. Several orbits are shown in Figure 8.4.

Points move along the orbits toward the origin. The directions of the eigenvectors X_1 and X_2 are indicated in the figure. Note that the orbit that begins at the point marked P approaches the origin first in the direction of X_2 and then in the direction X_1. This is because the X_2 component of the solution $X(t)$ decays most rapidly since λ_2 is the more negative eigenvalue.

PROBLEM 8.2

Consider the system (8.7) in the case that A is given by

$$A = \begin{bmatrix} 0 & 1 \\ 1 & 0 \end{bmatrix}.$$

This system arises in analyzing one of the singular points of the nonlinear pendulum system (Problem 8.3). Show that the origin is a saddle point for this linear system and sketch some of the orbits.

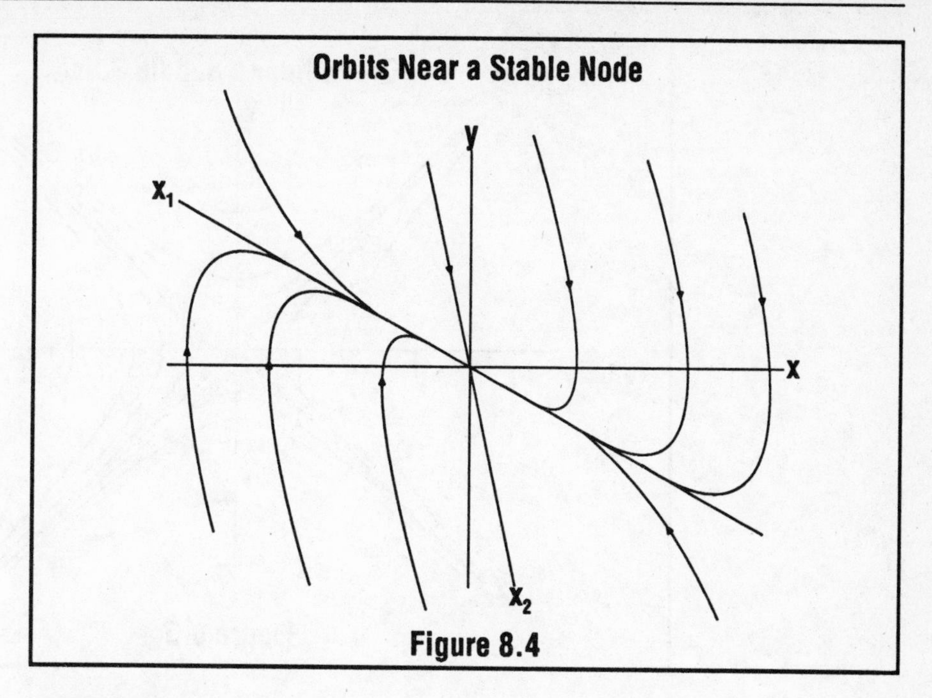

Figure 8.4

SOLUTION 8.2

The eigenvalues of the matrix A are easily found to be $\lambda_1 = -1$ and $\lambda_2 = 1$ with corresponding eigenvectors

$$X_1 = [1-1]^T \text{ and } X_2 = [1,1]^T.$$

Since the eigenvalues are real and of opposite sign, the origin is a *saddle point* for this system. The solution to the system is given by equation (8.8)(a) from which it is clear that points will move along orbits *toward* the origin along the direction X_1, the eigenvector associated with the negative eigenvalue –1. Points will move *away* from the origin along the direction X_2, which is associated with the positive eigenvalue λ_2. Several orbits are pictured in Figure 8.5 on which the directions X_1 and X_2 are indicated.

Since every orbit appears to escape eventually to infinity, a saddle point is an example of an unstable critical point.

Phase Plane Analysis of Nonlinear Systems

PROBLEM 8.3

In Problem 3.10 we showed that the angular displacement ϑ of a mass m on a frictionless pendulum of length L is governed by the equation

$$mL\,\vartheta''(t) = -mg \operatorname{Sin} \vartheta(t). \tag{1}$$

If we let $\varphi(t) = \vartheta'(t)$, then we can replace this equation by the first-order autonomous system

$$\vartheta'(t) = \varphi(t)$$

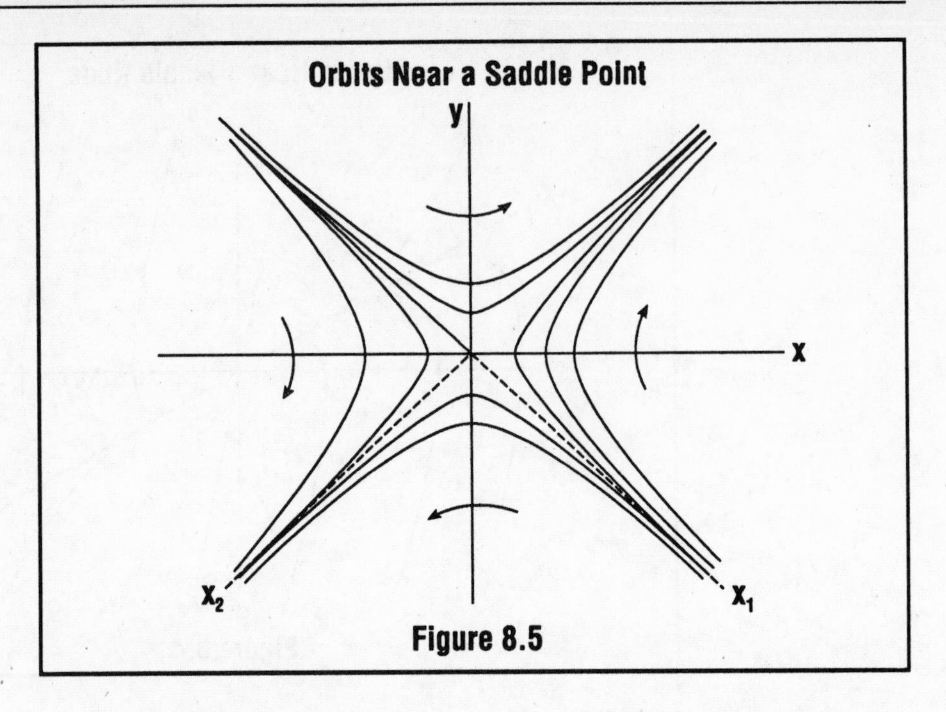

Figure 8.5

$$\varphi'(t) = -\omega^2 \text{Sin}\vartheta(t)$$

where $\omega^2 = g/L$. Find and classify all the critical points of this system. Sketch the orbits around the critical points and "fill in the blanks" to get an approximate global picture of the phase plane portrait for the nonlinear system.

SOLUTION 8.3

For this system we have

$$P(\vartheta,\varphi) = \varphi \quad \text{and} \quad Q(\vartheta,\varphi) = -\omega^2 \text{Sin } \vartheta, \tag{3}$$

from which it follows that $P = Q = 0$ if and only if

$$\varphi = 0 \quad \text{and} \quad \vartheta = k\pi, \quad k = \text{integer}. \tag{4}$$

Thus, (2) has *infinitely many* critical points located at integer multiples of π along the ϑ axis (horizontal axis). While linear systems have only one critical point, nonlinear systems may have any number of them (from zero to infinity).

Computing the partial derivatives of P and Q with respect to ϑ and φ, we find

At the critical points $(k\pi,0)$ this matrix reduces to

$$J(\vartheta,\varphi) = \begin{bmatrix} 0 & 1 \\ -\omega^2 \text{Cos}\vartheta & 0 \end{bmatrix} \tag{5}$$

and we find, then, the following eigenvalues for the matrix $J(k\pi,0)$:

$$J(k\pi,0) = \begin{bmatrix} 0 & 1 \\ -\omega^2(-1)^k & 0 \end{bmatrix} \tag{6}$$

$$\text{for } k = \text{even}, \quad \lambda_{1,2} = \pm i\omega$$
$$\text{for } k = \text{odd}, \quad \lambda_{1,2} = \pm \omega. \tag{7}$$

Thus, the critical point $(k\pi,0)$ is a *center* when k is even and a *saddle point* when k is an odd integer.

When k is odd, the eigenvectors corresponding the eigenvalues $\lambda_1 = \omega$ and $\lambda^2 = -\omega$ are found to be $X_1 = [1,\omega]^T$ and $X_2 = [1,-\omega]^T$. Then orbits approach the saddle points from the direction X_2 and depart in the direction X_1. The saddle points are unstable critical points for the linearized system and, by Theorem 8.1, they are unstable for the nonlinear system as well.

The stability of the critical points corresponding to even values of k is not determined by the theorem. Instead, we multiply both sides of the equation (1) by ϑ' and note that

$$d/dt(\vartheta'(t)^2/2) = \vartheta'(t)\vartheta''(t)$$

$$d/dt(1-\text{Cos}\vartheta(t)) = \vartheta'(t)\text{Sin}\vartheta(t).$$

Then (1) implies that for (ϑ,φ) a solution for (2), we have

$$E(\vartheta,\varphi) = 1/2\ \varphi(t)^2 + \omega^2(1-\text{Cos}\ \vartheta(t)) = \text{Constant}. \tag{8}$$

The quantity E is related to the total energy of the frictionless pendulum, and (8) states that the energy is constant on orbits. More precisely, the kinetic energy of the pendulum is proportional to $1/2\varphi(t)^2$, and the potential energy due to position is proportional to $\omega^2(1-\text{Cos}\vartheta)$. So E is a first integral for the system (2) and the level curves of the function $E(\vartheta,\varphi)$ are the orbits for (2) near the critical points. Since these orbits are closed curves, the critical points are centers.

We can examine the orbits about the center at the origin for various values of the constant E. First, $E = 0$ implies that $\vartheta(t) = \varphi(t) = 0$ for all t; i.e., the orbit corresponding to zero energy consists of the pendulum remaining motionless in the straight-down equilibrium position.

For small values of E, ϑ is small and it follows that $1-\text{Cos}\ \vartheta \approx \vartheta^2/2$. Thus for small values of E, we have

$$E \approx (\varphi^2 + \omega^2\vartheta^2)/2$$

and the level curves of this function are ellipses; i.e., the small amplitude orbits are elliptical closed curves corresponding to periodic oscillations of the pendulum. As the magnitude of E increases, the maximum values of ϑ and φ on the curves $E(\vartheta,\varphi) = E$ also increase, and the approximation of $(1-\text{Cos}\vartheta)$ by $\vartheta^2/2$ becomes less and less accurate. The orbits becomes less and less like ellipses until finally, when E reaches the value $2\omega^2$, the corresponding orbit connects the saddle point at $(-\pi,0)$ to the saddle point at $(\pi,0)$. A second orbit with $E = 2\omega^2$ connects the saddle point at $(\pi,0)$ to the saddle point at $(-\pi,0)$. Since the saddle points are critical points, it cannot happen that a point moving along one of these orbits will pass *through* the critical point onto the other orbit.

For values of E exceeding $2\omega^2$, the orbits are no longer closed curves, and the motion of the pendulum is then not periodic. For each value of E greater than $2\omega^2$ there are two possible orbits, one on which $\varphi = \vartheta'$ is always positive and one on which ϑ' is steadily negative. These correspond to orbits on which the pendulum revolves about the pivot point in the clockwise or counterclockwise direction, respectively. Physically, we interpret this to mean that if a fricitionless pendulum is set in motion with sufficiently high energy, it will revolve about the pivot point in one direction for all time.

The phase plane portrait of the region $-2\pi < \vartheta < 2\pi$, $-4\omega^2 < \varphi < 4\omega^2$ is sketched in Figure 8.6. The orbits corresponding to $E = 2\omega^2$ are called *separatrices.* The orbits with $E < 2\omega^2$ are closed curves of more or less elliptic shape. These correspond to periodic solutions, while the orbits with $E > 2\omega^2$ are the wavelike curves running parallel to the horizontal axis; i.e., ϑ tends to plus or minus infinity according to whether φ is positive or negative. These two types of orbits are separated from one another by the separatrices. Since the separatrices pass through the saddle points at $(\pm\pi,0)$, any motion that begins at a point on one of these curves will end up at one of the saddle points after an infinite amount of time.

For example, the orbit that begins at the point $(0,2\omega^2)$ moves in the clockwise direction toward $(\pi,0)$. Since the velocity decreases toward zero as the critical point is approached, it can be shown that the critical point cannot be reached in a finite time. Similarly, an orbit beginning at $(0, -2\omega^2)$ would move toward $(-\pi, 0)$ but could not reach this point in finite time.

PROBLEM 8.4

Find and classify the critical points of the system

$$\vartheta'(t) = \varphi(t)$$

$$\varphi'(t) = -\omega^2 \operatorname{Sin} \vartheta(t) - 2b\varphi(t).$$

which governs the motion of a pendulum with friction in which the frictional force is proportional to $\vartheta' = \varphi$. Sketch the phase plane portrait in the region $-2\pi < \vartheta < 2\pi$, $-4\omega^2 < \vartheta < 4\omega^2$, assuming that $0 < b < \omega$.

SOLUTION 8.4

Proceeding as in the last problem, we find that the critical points for the system (1) are located at $(k\pi,0)$ for k = integer. Similarly, we compute

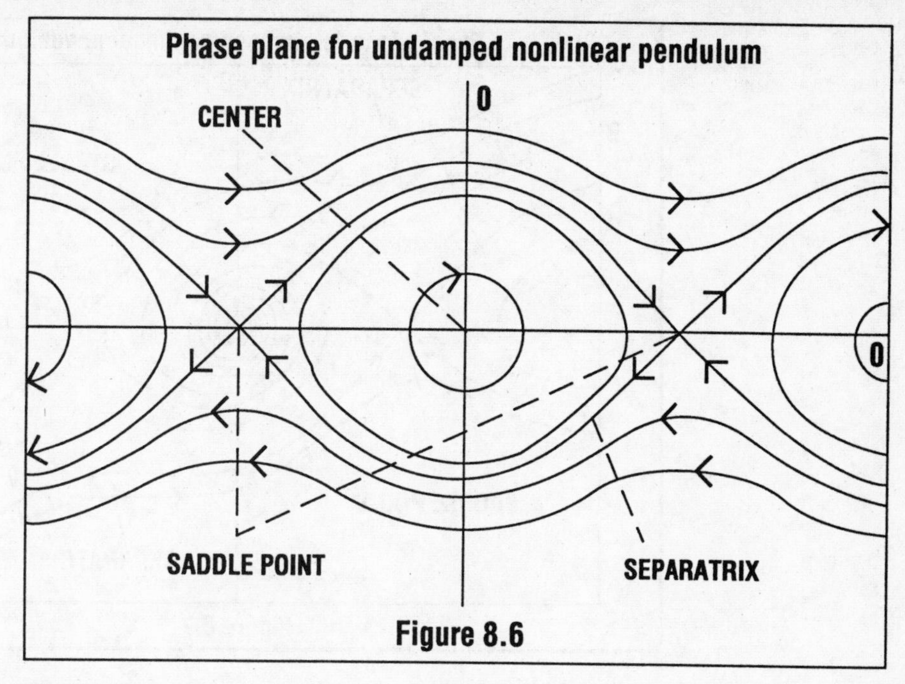

Figure 8.6

Then

and the eigenvalues of this matrix are the roots of the equation

$$\lambda(\lambda+2b) + (-1)^k \omega^2 = 0.$$

$$J(\vartheta,\varphi) = \begin{bmatrix} 0 & 1 \\ -\omega^2 \text{Cos}\vartheta & -2b \end{bmatrix}$$

Since we have assumed that $b^2 < \omega^2$, it follows that

$$J(k\pi,0) = \begin{bmatrix} 0 & 1 \\ -\omega^2(-1)^k & -2b \end{bmatrix},$$

if k = even $\quad \lambda_{1,2} = -b \pm \sqrt{b^2 - \omega^2} = -b \pm i\omega_D$

if k=odd $\quad \lambda_{1,2} = -b \pm \sqrt{b^2 - \omega^2}$

Then the critical point $(k\pi,0)$ is a stable focus when k is even and is a saddle point when k is odd. A sketch of the phase plane portrait in the region $-2\pi < \vartheta < 2\pi$ is shown in Figure 8.7.

Note that some orbits are attracted to the stable focus at the origin, while others approach one or the other of the stable foci at $(-2\pi,0)$ and $(2\pi,0)$. In particular, the orbit that starts at the point A is attracted eventually to the focus at $(2\pi,0)$, while the orbits that begin at points B and C are attracted to the origin and $(-2\pi,0)$, respectively.

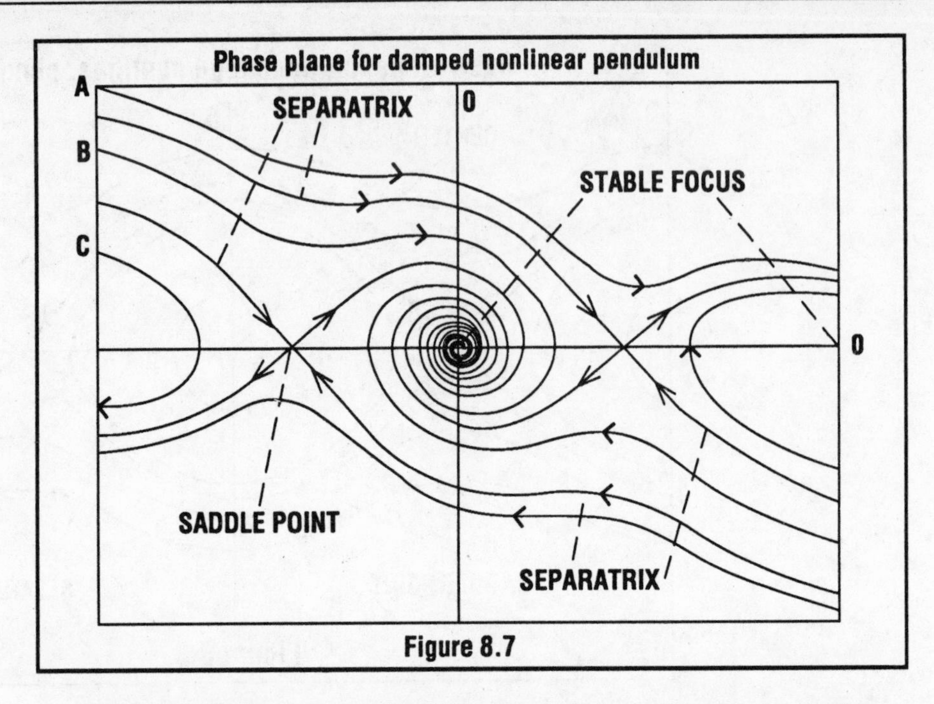

Figure 8.7

The orbit that begins at point A is separated from the orbit beginning at point B by the orbit that passes through the saddle point at $(\pi,0)$. This orbit, starting at a point between A and B and ending at the saddle point $(\pi,0)$, is a separatrix. Similarly, the orbit beginning at C is attracted to the stable focus at $(-2\pi,0)$, and this orbit is separated from the orbit beginning at B by another separatrix. This separatrix begins at a point between B and C and ends at the saddle point at $(-\pi,0)$. Any orbit that begins at a point lying *between* these two separatrices will eventually spiral into the stable focus at the origin. This region is referred to as the *basin of attraction* for the stable focus at the origin.

Neutrally Stable Critical Points

PROBLEM 8.5

Show that the origin is the only critical point for the system

$$\begin{aligned} x'(t) &= -y - \beta x^3 \\ y'(t) &= x. \end{aligned} \tag{1}$$

Show that the origin is a stable focus if the parameter β is positive.

SOLUTION 8.5

Clearly the origin is the only point in the phase plane where the functions $P(x,y) -y - \beta x^3$ and $Q(x,y) = x$ vanish simultaneously. Since
it is easy to see that the eigenvalues of $J(0,0)$ are $+i$ and $-i$, which implies that the origin is a *center* for the *linearization about the origin* for system (1). It does not necessarily follow that the origin is a center for the nonlinear

$$J(x,y) = \begin{bmatrix} -3\beta x^2 & -1 \\ 1 & 0 \end{bmatrix},$$

system. The failure of the linear approximation to shed light on the nature of the orbits of the nonlinear system near a center is not hard to explain.

Near an asymptotically stable critical point of a nonlinear system, the orbits of the approximating linearized system are inward spirals, or they are curves that are roughly hyperbolic in shape. Near an unstable critical point the orbits have much the same shape, but points move away from the critical point along the orbits instead of toward the critical point. The orbits of the nonlinear system must lie close to the orbits of the approximating linear system near the critical point, and a slightly perturbed spiral is still a spiral (more or less). The same may be said of a curve that is hyperbolic in shape. However, if the critical point is a center for the linear system, then the orbits of the linear system are *closed* curves. Perturbing a closed curve even slightly may produce a curve that is no longer closed, hence the orbits of the nonlinear system may be any one of the following: inward spirals, closed curves, or outward spirals. It may be possible to use a version of the following trick to determine which of these cases applies.

Multiplying the first equation in system (1) by 2x and the second by 2y and adding the resulting equations leads to

$$d/dt(x^2+y^2) = -2\beta x^4. \tag{2}$$

Here we have used that

$$2xx' + 2\,yy' = d/dt(x^2+y^2). \tag{3}$$

The equation (2) does not produce a first integral for the system; it does not identify a quantity that is constant along orbits. However, if we let $r^2=x^2+y^2$ denote the square of the distance from the origin to a point (x,y) on an orbit, then (2) implies that r^2 is everywhere decreasing if $\beta>0$. It follows that the orbits behave like inward spirals and the origin must be a stable focus for the system (1). The observation (3) is often (but certainly not always) helpful in determining the stability of a critical point that is neutrally stable for the linearized system.

PROBLEM 8.6

Compute the linearization about the critical points for the system,

$$\begin{aligned} x'(t) &= (xy)^4 \\ y'(t) &= x^4-y^4 \end{aligned} \tag{1}$$

SOLUTION 8.6

The origin is easily seen to be the only critical point for this system. We compute

$$J(x,y) = \begin{bmatrix} 4y(xy)^3 & 4x(xy)^3 \\ 4x^3 & -4y^3 \end{bmatrix}$$

Then J(0,0) is the 2 by 2 matrix of all zeroes, and the linearization of the nonlinear system about the critical point is the trivial system. The linear system provides no information about the behavior of the nonlinear system near the critical point. This example illustrates that while the technique of linearization is effective in many cases, it is not a universally successful method for dealing with nonlinear problems.

PROBLEM 8.7

In Chapter 3 we discussed models for single species population growth. A two species population model in which the population denoted by Y(t) preys upon the population denoted by X(t) is described by the system of equations

$$\begin{aligned} X'(t) &= (b-\delta Y(t))X(t) \\ Y'(t) &= (\beta X(t)-d)Y(t). \end{aligned} \tag{1}$$

The equations in (1) express the fact that population X, the prey, has a constant birth rate b, but the death rate is directly proportional to the size of the predator population. Similarly, the Y population, the predators, has a constant death rate d, but the birth rate is directly proportional to the size of the prey population. Roughly speaking, this amounts to saying that there is an unlimited food supply for the prey population and that there is no population that is preying on the predators. Find all the equilibrium solutions to this system and determine whether any of them is stable.

SOLUTION 8.7

Note that if the predator population Y equals zero, then the prey population grows exponentially without bound. Similarly, if the prey population X equals zero, then the predator population decreases exponentially to zero. We are interested in determining whether there are any stable solutions other than these two extremes. In particular, we want to know if it is possible for the two populations to exist indefinitely without either of them dying out.

The equations

$$(b - \delta Y(t))X(t) = 0$$

$$(\beta X(t)-d)\, Y(t) = 0$$

are satisfied for X = Y = 0 and X = d/β, Y = b/δ. We compute

$$J(X,Y) = \begin{bmatrix} b-\delta Y & -\delta Y \\ \beta Y & \beta X-d \end{bmatrix}.$$

Then

$$J(0,0) = \begin{bmatrix} b & 0 \\ 0 & -d \end{bmatrix}$$

and since the eigenvalues of this matrix are equal to $\lambda_{1,2}$ = b,–d, it follows that 1,2 the origin is a saddle point for this system. Note that the eigenvector corresponding to $\lambda_1 = b$ is the vector $e_1 = [1,0]^T$ and $e_2 = [0,1]^T$ is the eigenvector associated with the negative eigenvalue $\lambda_2 = -d$. Thus, if the predator population Y is zero then the prey population grows to infinity along e_1 while if the prey population X is zero then the predator population decreases to zero along e_2. So (0,0) is an unstable, saddle-type equilibrium point.

Since negative values for either population have no physical significance, we restrict our attention to the first quadrant of the phase plane. The X and Y axes are orbits for this system, and since distinct orbits never cross, no orbit that begins in the first quadrant can cross into one of the other quadrants.

Now

$$J(d/\beta,b/\delta) = \begin{bmatrix} 0 & -d\delta/\beta \\ b\beta/d & 0 \end{bmatrix}$$

has eigenvalues

$$\lambda_{1,2} = \pm\, i\sqrt{bd}$$

hence (d/β,b/δ) is a center for the linearization of (1) about this critical point. We must employ a version of the trick discussed in Problem 8.5 in order to determine the stability of this equilibrium point.

After some trial and error, we find that if we multiply the first equation in (1) by $(\beta X-d)/X$ and multiply both sides of the second equation by $(b-\delta Y)/Y$, then subtraction leads to

$$X'(t)(\beta-d/X(t))+Y'(t)(\delta-b/Y(t)) = 0\,. \tag{2}$$

This can be written as

$$d/dt(\beta X(t) - d\ \ln X(t) + \delta Y(t) - b\ln Y(t)) = 0\,. \tag{3}$$

Then along each orbit of the system (1), we have that

$$F(X,Y) = \beta X(t)+\delta Y(t) - \ln(X^d Y^b) = \text{Constant}; \tag{4}$$

i.e., (4) is a first integral for (1). While it is not easy to see what type of curves are the level curves for the function F(X,Y) in (4), we can sketch the phase plane plot in the first quadrant using numerical solution methods. Using a more delicate argument, we can even prove that the level curves of F are necessarily closed curves.

The phase plane portrait for this system is shown in Figure 8.8. The existence of closed orbits in the first quadrant implies that for predator-prey population interactions described by (1), the populations can vary periodically with each population surviving indefinitely. Note that the orbits in the first quadrant are not attracted to the critical point there. The equilibrium point in the first quadrant is stable but not asymptotically stable. The orbits in quadrants two, three, and four near the unstable critical point at the origin have no significance in the context of population models but have been sketched anyway to emphasize the saddle point nature of this critical point.

Limit Cycle Behavior

PROBLEM 8.8

A spring mass system with an unusual friction force is governed by

$$mx''(t) = -Kx(t) - m(px'(t) - 2d)x'(t), \tag{1}$$

where p and d denote positive constants. This friction force is unusual because it changes sign depending on the size of $x'(t)$. Equation (1) is equivalent to the autonomous system

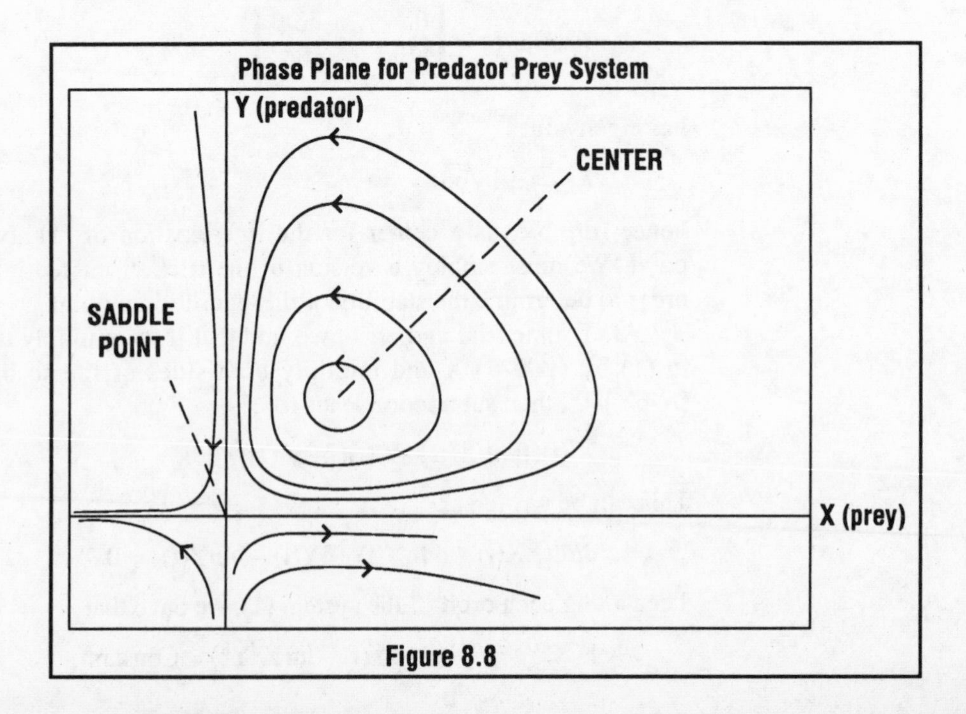

Figure 8.8

$$x'(t) = y(t)$$
$$y'(t) = -\omega^2 x(t) - (py^2 - 2d)y. \tag{2}$$

Show that the origin is the only critical point for this system and that the system possesses a limit cycle enclosing the origin.

SOLUTION 8.8

It is easy to see that x=y= 0 is the only point where

$$y = -\omega^2 x - y(py^2 - 2d) = 0.$$

We compute

$$J(0,0) = \begin{bmatrix} 0 & 1 \\ -\omega^2 & 2d \end{bmatrix}$$

for which the eigenvalues are

$$\lambda_{1,2} = d \pm \sqrt{d^2 - \omega^2}. \tag{3}$$

For d such that $0 < d < \omega$, the origin is an unstable focus; i.e., the eigenvalues are a complex conjugate pair with positive real part. Then the orbits spiral outward from the origin. Note, however, that if we multiply the first equation in (2) by 2kx and the second equation by 2y, then adding the result leads to

$$d/dt(kx^2 + y\ ) = -2(py^4 - dy^2). \tag{4}$$

Here we are proceeding as we did in Problem 8.5, and we have used the fact that

$$d/dt(kx^2 + y^2) = 2kxx'(t) + 2yy'(t).$$

It is apparent from (4) that for y^2 sufficiently large (i.e. for $y^2 > d/p$), the quantity $kx^2 + y^2$ is a decreasing function of time. Since $kx^2 + y^2$ is related to the distance of the point (x,y) from the origin, it follows that for y^2 large, the orbits behave like inward spirals. On the other hand, for small values of y^2 (i.e., for $y^2 < d/p$) it is also evident from (4) that $kx^2 + y^2$ is increasing with time.

This last observation is consistent with the fact that the origin is an unstable focus. However, the fact that orbits on which y^2 assumes large values are drawn back toward the origin makes it difficult to imagine how the phase plane portrait for this system must look.

Consider then the roughly annular region U defined by

$$d/2p < kx^2 + y^2 < 2d/p.$$

The region U excludes the "hole" where $kx^2 + y^2 \leq d/2p$. In particular, U excludes the origin so that U contains no critical points of the system (2).

Equation (4) implies that orbits originating in the "hole" are forced to spiral outward away from the origin. Similarly, (4) implies that orbits that originate outside of U, where $kx^2 + y^2 \geq 2d/p$, are attracted back toward the origin and toward U. This does not *prove* that orbits that originate in U remain inside U, but it is suggestive that Theorem 8.2, the Poincare-Bendixson theorem, may be applied in order to conclude that the region U contains a limit cycle. The phase plane portrait for this system is sketched in Figure 8.9. Since the limit cycle to which the orbits are attracted is a closed curve, the solutions of this system tend with time to periodic solutions.

Note that when the parameter d is negative, then the critical point at (0,0) is an asymptotically stable focus; all orbits are attracted to the origin. For d positive, we have just shown that the origin is an unstable focus and the system has a limit cycle attractor. That is, for $d < 0$ all solutions tend to zero, while for $d > 0$ all solutions tend to a periodic solution. In the context of the physical system, this may be interpreted to mean that when $d < 0$ every initial state produces a motion that is oscillatory with amplitude decaying steadily to zero and when $d > 0$ there are initial states that produce a motion that tends to a self-sustaining periodic motion.

This transition from solutions that exhibit one sort of behavior (i.e., solutions that decay to zero) to solutions exhibiting a completely different type of behavior, (i.e. periodic solutions) as the parameter d changes from negative to positive is an example of the phenomenon of *bifurcation*. Bifurcation is a purely nonlinear phenomenon in which a system experiences a dramatic change in behavior as a parameter, called the *bifurcation parame-*

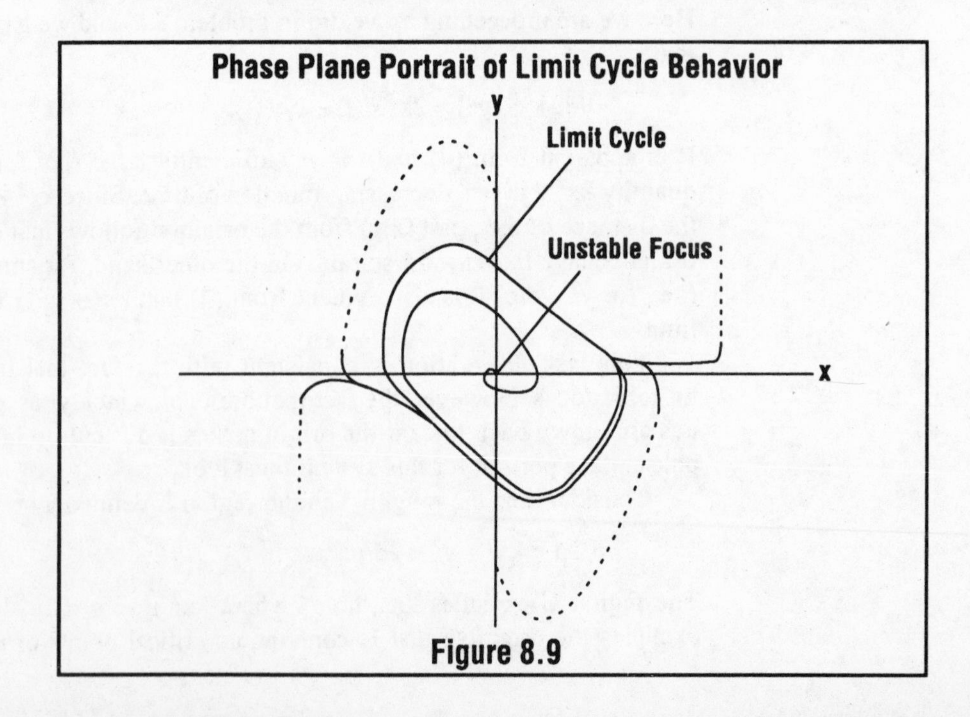

Figure 8.9

ter, passes through a critical value called a *bifurcation point*. Here the parameter d is the bifurcation parameter, and the bifurcation point is d = 0. Analysis of bifurcation phenomena involves methods that are beyond the scope of this text.

Systems of nonlinear differential equations arise very frequently in applications. In general these problems are not explicitly solvable but it is often possible to discover considerable information about the behavior of the solution. This type of analysis is usually feasible only for systems that are autonomous; i.e., for equations that do not explicitly involve the independent variable. For autonomous systems it is possible to examine the behavior of the solution in the vicinity of certain special points called critical points and draw conclusions about the global behavior of the solution.

This procedure is theoretically possible for a system of any size, but for a system of two equations in two unknown functions x and y, it is possible to present the results visually by plotting solution curves in the xy plane. We refer to the xy plane as the phase plane and to the solution curves as orbits.

The local analysis of a nonlinear system is carried out as follows. First, the critical points are identified by finding all points where the forcing terms in the equations simultaneously vanish and $x'(t) = y'(t) = 0$. *Critical points are also called* fixed points *or* equilibrium points.

Often, but not always, the orbits of a nonlinear system can be approximated in the vicinity of a critical point by the orbits of an associated linear *system. The character of the orbits for the linear system is easily determined by computing the eigenvalues of the coefficient matrix for the linear system. Once the local behavior of the orbits near each of the critical points is known, it should be possible to piece together the local pictures to get an idea of the global phase plane portrait for the nonlinear system.*

Although a linear system has a single critical point at the origin, a nonlinear system may have any number of critical points, and they may be located anywhere in the phase plane. Each critical point may be classified as stable, asymptotically stable, or unstable. A critical point for a nonlinear system is asymptotically stable (unstable) if the origin is asymptotically stable (unstable) for the associated linear system. If the origin is stable but not asymptotically stable as a critical point of the linearized system, then more delicate methods must be used to determine the stability or instability of the critical point for the nonlinear system. For example, a quantity that is constant on orbits of the system is called a first integral *of the system, and the stability of a critical point is generally evident from a first integral. However, it is not always possible to find first integrals for nonlinear systems.*

An asymptotically stable fixed point attracts nearby orbits and we say that this fixed point is an attractor. *A two dimensional nonlinear system may have attracting fixed point(s) or it may have a simple closed curve attractor; such curves are called limit cycles and do not occur for linear systems. For nonlinear systems of dimension higher than two, more complicated attractors are possible. Such systems may even produce attractors whose dimension is not equal to an integer; these are called* strange attractors *and the associated behavior of the dynamical system is referred to as* chaotic *behavior.*

9

Power Series Methods

The solution methods presented in previous chapters are limited primarily to linear equations having constant coefficients. We know that the problem of finding a solution for a linear differential equation with constant coefficients can be reduced to the algebraic problem of finding the roots of the auxiliary equation. This reduction may be accomplished by assuming the solution is a function of exponential form.

Equations with variable coefficients cannot be reduced to equivalent algebraic problems by such a simple assumption. However, assuming that the solution can be expanded in a power series may lead to an algebraic recursion relation for the unknown coefficients in the power series. The power series then provides a representation of the solution in a neighborhood of the expansion point. The method is illustrated in this chapter for equations of the form

$$p(x)y''(x) + q(x)y'(x) + r(x)y(x) = 0$$

i.e., homogeneous, linear equations of the second order having coefficients that are polynomials. Although the method applies more generally to homogeneous and inhomogeneous linear equations of higher order having coefficients that are more general than polynomials., it will be illustrated here only for the case of polynomial coefficients.

TERMINOLOGY

ORDINARY POINTS AND SINGULAR POINTS

A point x_0 is called an *ordinary point* for the differential equation above if $p(x_0)\neq 0$. If x_0 is not an ordinary point, then it is called a *singular point.* The power series method involving expansions about an ordinary point is straightforward. This technique, applied to a linear equation of order N, leads invariably to N linearly independent solutions with power series representations. The method is illustrated in this chapter for equations of order two.

THE METHOD OF FROBENIUS

Singular points may be further classified as *regular singular points* and *irregular singular points.* The precise definition of a regular singular point is given later. A singular point that is not regular is said to be irregular. The power series method, modified to permit expansion of the solution about a regular singular point is called the *Frobenius* method, which we describe in this chapter.There is no general theory for expanding solutions about irregular singular points, so we will not discuss solutions near irregular singular points.

Before discussing the expansion of solutions in power series, we shall recall a few of the fundamental facts about infinite series of functions in general, and power series in particular.

INFINITE SERIES

Let $u_0(x),u_1(x), ...,u_n(x)$ denote a family of functions having a common domain of definition D. Then at any point x in D the expression

$$u_0(x)+u_1(x)+ ... = \sum_{n=0}^{\infty} u_n(x) \tag{9.1}$$

is called an *infnite series* of functions.

Convergence

The series (9.1) is said to converge at x_0 in D if

$$\lim_{N\to\infty} \sum_{n=0}^{N} u_n(x_0) \quad \text{exists;} \tag{9.2}$$

i.e., the series (9.1) converges at x_0 in D if

for every $\varepsilon > 0$ there exists a positive number M such that

$$\left|\sum_{n=p}^{q} u_n(x_0)\right| < \varepsilon \qquad \text{for all p and q such that } p, q > M \tag{9.3}$$

The limit in (9.2) exists if the terms at the "tail end" of the series are of diminishing importance. The statement (9.3) says this in a precise way.

ABSOLUTE CONVERGENCE

The series (9.1) is said to be *absolutely convergent* at x_0 if the series

$$\sum_{n=0}^{\infty} |u_n(x)| \tag{9.4}$$

is convergent at x_0. Clearly a series is convergent if it is absolutely convergent, but the converse is false. However, it is easier to test a series for *absolute* convergence since it involves series of positive terms with no possibility of cancellations.

For example, we have the following theorem.

Theorem 9.1 The Ratio Test

Theorem 9.1 The series (9.1) is absolutely convergent at x_0 in D if

$$\lim_{n\to\infty} |u_{n+1}(x_0)/u_n(x_0)| < 1\,. \tag{9.5}$$

Example 9.1 The Ratio Test

9.1(A)

Consider the series

$$\sum_{n=0}^{\infty} (x/2)^n\,. \tag{9.6}$$

Then

$$|u_{n+1}(x_0)/u_n(x_0)| = |x/2|$$

and the condition (9.5) holds at each point x_0 where $|x_0/2| < 1$; i.e. for $-2 < x_0 < 2$. According to Theorem 9.1, this series is absolutely convergent at x_0 for all x_0 in the interval $(-2,2)$.

9.1(B)

For the series

$$\sum_{n=0}^{\infty} (x-1)^n/n! \tag{9.7}$$

we have

$$|u_{n+1}(x_0)/u_n(x_0)| = |(x_0-1)/(n+1)|$$

As n tends to infinity, this expression tends to zero for each fixed x_0. We conclude that the series (9.7) is absolutely convergent at x_0 for *every* x_0.

UNIFORM CONVERGENCE

Let I denote an open interval contained in D. The series (9.1) is said to be *uniformly convergent* on I if the condition (9.3) holds at each point x_0 in I and the parameter M depends on ε but does *not* depend on x_0. We have then the following theorem.

Theorem 9.2 Weierstrass M-test

Theorem 9.2 The series (9.1) is absolutely and uniformly convergent on an interval I if

(a) there exists a sequence of positive constants M_n such that for all x in I and every integer n

$$|u_n(x)| \le M_n \qquad \text{and}$$

(b) $\sum_{n=0}^{\infty} M_n$ converges.

Example 9.2 Uniform Convergence

9.2(A)

The series in (9.6) converges absolutely for all x such that $|x| < 2$. Then for a fixed positive value $a < 2$, let I be the open interval $(-a,a)$. It follows from Example 9.1(a) that the constants $M_n = (a/2)^n$ satisfy condition (b) of Theorem 9.1. In addition, for each n,

$$|x/2|^n < M_n = (a/2)^n \qquad \text{for all x in } I = (-a,a).$$

It follows then from Theorem 9.2 that the series (9.6) is uniformly convergent on the interval $(-a,a)$ for every $a < 2$.

9.2(B)

Let R denote a positive number and let I denote the interval $\{|x-1\}| < R\}$. It follows from part (b) of example 9.1 that the positive constants $M_n = R^n/n!$ satisfy condition (b) of Theorem 9.2. In addition,

$$|(x-1)^n/n!| \le M_n = R^n/n! \qquad \text{for all x in } I = \{|x-1| < R\}$$

It follows from Theorem 9.2 that the series (9.7) converges uniformly on the interval I. Note that the constants M_n satisfy condition (b) of Theorem 9.1 for all positive values R. Hence the series (9.7) is uniformly convergent on I for every positive R, which is to say on every interval of the real line.

Power Series

If the family of functions in (9.1) is given by

$$u_n(x) = a_n (x-x_0)^n \quad n = 0,1, \ldots$$

for given constants a_n, we say that the series (9.1) is a *power series* about the point x_0. We state now, without proof, several facts about power series.

PROPERTIES OF POWER SERIES

1. Convergence of Power Series The power series

$$\sum_{n=0}^{\infty} a_n (x-x_0)^n \tag{9.8}$$

converges absolutely and uniformly on an interval $I = \{|x-x_0| < R\}$. The number R is called the *radius of convergence* for the series. R may be zero, positive, or infinite, depending on the coefficients a_n in the series. Note that although power series converge absolutely and uniformly on the interval of convergence, these properties are not related for series in general. There are infinite series that converge absolutely that do not converge uniformly and vice versa.

2. Differentiation of Power Series The power series (9.8) defines a function F(x) whose domain is the interval I. The series

$$\sum_{n=1}^{\infty} n a_n (x-x_0)^{n-1}$$

obtained from (9.8) by differentiation, converges absolutely and uniformly on I to the function $F'(x)$. We may differentiate (9.8) as often as we like. The resulting differentiated series converges absolutely and uniformly on I to the corresponding derivative of F(x).

3. Integration of Power Series The power series (9.8) may be integrated term by term to obtain the series

$$\sum_{n=0}^{\infty} a_n (x-x_0)^{n+1}/(n+1)$$

which converges absolutely and uniformly on I to G(x), where $G'(x)=F(x)$ and $G(x_0)=0$.

4. Sums and Products of Power Series Suppose the power series

$$\sum_{n=0}^{\infty} a_n (x-x_0)^n \quad \text{and} \quad \sum_{n=0}^{\infty} b_n (x-x_0)^n$$

each converge on the interval $I= \{|x-x_0| < R\}$. Then each of the following series also converges absolutely and uniformly on I:

$$\sum_{n=0}^{\infty} a_n (x-x_0) \pm \sum_{n=0}^{\infty} b_n (x-x_0)^n = \sum_{n=0}^{\infty} (a_n \pm b_n)(x-x_0)^n$$

$$\sum_{n=0}^{\infty} a_n (x-x_0) \sum_{n=0}^{\infty} b_n (x-x_0)^n = \sum_{n=0}^{\infty} c_n (x-x_0)^n$$

where $c_n = a_0 b_n + a_1 b_{n-1} + \ldots + a_n b_0$

Thus, convergent power series may be added, subtracted, and multiplied.

5. Equality of Power Series Suppose that two series, each convergent on the interval I, satisfy

$$\sum_{n=0}^{\infty} a_n (x-x_0)^n = \sum_{n=0}^{\infty} b_n (x-x_0)^n \qquad \text{for all x in I.}$$

Then, $a_n = b_n$ for all n.

Taylor Series for Analytic Functions

Note that if the series (9.8) has a positive radius of convergence, and if the function defined by the series is denoted by F(x), then

$$F^{(n)}(x_0) = n!\, a_n \qquad \text{for } n=0,1, \ldots$$

Conversely, given a function F(x) and a point x_0, consider the series

$$\sum_{n=0}^{\infty} F^{(n)}(x_0)(x-x_0)^n/n! \tag{9.9}$$

If the series (9.9) has a positive radius of convergence then the function F(x) is said to be *analytic* at x_0. The series (9.9) is called the *Taylor series* for F(x) at x_0.

Example 9.3 Analytic Functions

9.3(A)

Every polynomial in x, of any degree, is analytic at x_0 for every point x_0. Moreover, for any polynomial the radius of convergence for the Taylor series about every point is equal to infinity. For example, consider

$$F(x) = x^2 + 3x + 2.$$

Then $F(1) = 6$, $F'(1) = 5$, $F''(1) = 2$, $F^{(n)}(1) = 0$ for $n > 2$, and the Taylor series is equal to

$$F(x) = 6 + 5(x-1) + 2(x-1)^2/2! + 0 = 2 + 3x + x^2.$$

Since the Taylor series has only finitely many (three) terms, it is clearly convergent for all values of x.

9.3(B)

Each of the functions e^x, Sin x, and Cos x is analytic at every point x_0, and the Taylor series for any of these functions about any point x_0 has radius of convergence equal to infinity. For example, for $F(x)=e^x$

$$F^{(n)}(x) = e^x \text{ for all n} \quad \text{and} \quad F^{(n)}(0) = 1 \text{ for all n.}$$

Then the Taylor series is

$$e^x = \sum_{n=0}^{\infty} x^n/n!\,,$$

and the ratio test shows that this series converges absolutely for all values of x. Similarly, for the function F(x)=Cosh (2x), we have

$$F^{(2n)}(0) = 2^{2n}, \quad F^{(2n+1)}(0) = 0 \quad n = 0,1, \ldots$$

hence

$$\text{Cosh}\,(2x) = \sum_{n=0}^{\infty} (2x)^{2n}/(2n)!\,.$$

Differentiating this everywhere convergent series, term by term, leads to

$$2\,\text{Sinh}\,(2x) = \sum_{n=0}^{\infty} 2n\,(2x)^{2n-1}\,2/(2n)!$$

Thus the series for Sinh 2x is given by

$$\text{Sinh}\,(2x) = \sum_{n=0}^{\infty} (2x)^{2n-1}/(2n\text{-}1)!$$

This series has the same radius of convergence as the series for Cosh 2x, namely $R = \infty$.

POWER SERIES SOLUTIONS FOR DIFFERENTIAL EQUATIONS

Expansion About An Ordinary Point

Consider the linear, homogeneous, second order equation

$$p(x)y''(x) + q(x)\,y'(x) + r(x)\,y(x) = 0. \tag{9.10}$$

Dividing through by p(x) reduces the equation to *normal form*

$$y''(x) + A(x)y'(x) + B(x)\,y(x) = 0. \tag{9.11}$$

The point x_0 is said to be an *ordinary point* for (9.10) (and for (9.11)) if both A(x) and B(x) have convergent power series expansion about x_0 and these series have positive radii of convergence. If either A(x) or B(x) fails to have such an expansion, then x_0 is said to be a *singular point* for the equation. In the special case that p(x), q(x), and r(x) are polynomials in x having no factors in common, the point x_0 is an ordinary point if $p(x_0) \neq 0$, and it is a singular point if $p(x_0) = 0$.

Theorem 9.3 Expansion About an Ordinary Point

Theorem 9.3 If x_0 is an ordinary point for the differential equation (9.10), then the general solution is given by

$$y(x) = \sum_{n=0}^{\infty} a_n (x - x_0)^n = a_0 Y_0 (x) + a_1 Y_1 (x)$$

for arbitrary constants a_0 and a_1. The power series for $Y_0(x)$ and $Y_1(x)$ converge on the interval $|x - x_0| < R$, where R is the distance from x_0 to the nearest singular point of the differential equation.

Example 9.4 Expansion About An Ordinary Point

Consider the equation

$$y''(x) - 4y(x) = 0. \tag{9.12}$$

This equation is linear with constant coefficients and can be solved by the methods introduced in Chapter 4. However, for purposes of illustration we shall solve it by expanding about the ordinary point $x_0 = 0$. Since the equation has no singular points, Theorem 9.3 tells us that the series we develop are going to have radii of convergence equal to infinity.

We suppose (9.12) has a solution of the form

$$y(x) = \sum_{n=0}^{\infty} a_n x^n . \tag{9.13}$$

Then

$$y''(x) = \sum_{n=0}^{\infty} n(n-1)\, a_n x^{n-2},$$

and

$$\sum_{n=2}^{\infty} n(n-1)\, a_n x^{n-2} - \sum_{n=0}^{\infty} 4 a_n x^n = 0 . \tag{9.14}$$

Shifting the index Note that the sum

$$\sum_{n=2}^{\infty} n(n-1)\, a_n x^{n-2} = 2a_2 + 6a_3 x + 12a_4 x^2 + \ldots$$

and the sum

$$\sum_{n=0}^{\infty} (n+2)(n+1)\, a_{n+2} x^n = 2a_2 + 6a_3 x + 12a_4 x^2 + \ldots$$

produce the same series of terms. Replacing the index n everywhere in the first sum by n+2 leads to the second sum. Using this shifted sum in (9.14) produces

$$\sum_{n=0}^{\infty} (n+2)(n+1)\, a_{n+2}x^n - \sum_{n=0}^{\infty} 4a_n x_n = 0, \tag{9.15}$$

The Recursion Relation It follows from (9.15) and Property 5 of power series that

$$a_{n+2} = \frac{4}{(n+2)(n+1)}\, a_n \qquad \text{for } n=0,1,\ldots \tag{9.16}$$

Equation (9.16) is called the *recursion relation* for the coefficients in the solution to (9.12). Applying (9.16) for successive values of n produces

$$a_2 = \frac{4}{2 \cdot 1}\, a_0 \qquad\qquad a_3 = \frac{4}{3 \cdot 2}\, a_1$$

$$a_4 = \frac{4}{4 \cdot 3}\, a_2 = \frac{4^2}{4!}\, a_0 \qquad\qquad a_5 = \frac{4}{4 \cdot 3}\, a_3 = \frac{4^2}{5!}\, a_1$$

$$a_6 = \frac{4^3}{6!}\, a_0 \qquad\qquad a_7 = \frac{4^3}{7!}\, a_1 \ldots$$

Using these results in the series (9.13) then leads to the following solution for (9.12).

$$y(x) = a_0 \left[1 + \frac{4x^2}{2!} + \frac{4^2x^4}{4!} + \frac{4^3x^6}{6!} + \ldots\right]$$

$$+ a_1 \left[x + \frac{4x^3}{3!} + \frac{4^2x^5}{5!} + \frac{4^3x^7}{7!} + \ldots\right]$$

$$= a_0 \left[1 + \frac{(2x)^2}{2!} + \frac{(2x)^4}{4!} + \frac{(2x)^6}{6!} + \ldots\right]$$

$$+ 1/2\, a_1 \left[2x + \frac{(2x)^3}{3!} + \frac{(2x)^5}{5!} + \frac{(2x)^7}{7!} + \ldots\right]$$

Recalling example 9.2(b), we recognize these series as the series for Cosh 2x and Sinh 2x; i.e.

$$y(x) = a_0 \text{ Cosh } 2x + 1/2\, a_1 \text{ Sinh } 2x .$$

Note that when the equation (9.12) is written in the form (9.14), it is not easy to equate coefficients of like powers of x. However, when the index is shifted in the first sum and the equation is rewritten in the form (9.15), equating coefficients is easily carried out, leading to the recursion relation (9.16).

Expansion About A Singular Point

SINGULAR POINTS

If the point x_0 is not an ordinary point for the equation (9.11), then it is said to be a *singular point*. Singular points are classified further as *regular singular points* and *irregular singular points*. The point x_0 is a *regular singular point* for the equation (9.11) if both of the functions

$$(x - x_0)A(x) \quad \text{and} \quad (x - x_0)^2 B(x)$$

are analytic at x_0: i.e., they have convergent power series with positive radii of convergence about the point x_0. In the special case that p(x), q(x), and r(x) in (9.10) are all polynomials, then we may say equivalently that x_0 is a regular singular point for (9.10) if $p(x_0) = 0$ and each of the following limits exists:

$$\lim_{x \to x0} (x - x_0)\frac{q(x)}{p(x)} \quad \text{and} \quad \lim_{x \to x0} (x - x_0)^2 \frac{r(x)}{p(x)} .$$

If x_0 is not a regular singular point, then it is said to be an *irregular singular point*.

Example 9.5 The Euler Equation

Consider the equation

$$x^2 y''(x) + axy'(x) + by(x) = 0, \qquad x>0, \tag{9.17}$$

where a and b denote real constants. Since $p(x) = x^2$ vanishes at x=0, it is evident that x=0 is a singular point. Note that both of the functions

$$x\frac{q(x)}{p(x)} = x\frac{ax}{x^2} \quad \text{and} \quad x^2\frac{r(x)}{p(x)} = x^2\frac{b}{x^2}$$

have a finite limit as x tends to zero. Then x = 0 is a regular singular point for equation (9.17), which is known as *Euler's equation*. If we suppose that (9.17) has a solution of the form

$$y(x) = x^r, \tag{9.18}$$

then substituting (9.18) into (9.17) leads to the observation that the equation (9.17) is satisfied for all x > 0, if r satisfies

$$P(r) = r(r - 1) + ar + b = 0. \tag{9.19}$$

Since (9.19) is a quadratic equation in the unknown r, there are three possible situations with regard to the roots and the corresponding solutions:

1. **Real, distinct roots:** $P(r) = (r - r_1)(r - r_2)$
   In this case the general solution of (9.17) is of the form

$$y(x) = C_1 x^{r_1} + C_2 x^{r_2} \tag{9.20}$$

2. **Repeated roots:** $P(r) = (r - r_1)^2$
   In this case the general solution of (9.17) is of the form

$$y(x) = [C_1 + C_2 \ln x]x^{r_1} \tag{9.21}$$

3. Complex conjugate roots: $r_{1,2} = \lambda \pm i\mu$
In this case the general solution of (9.17) is of the form

$$y(x) = [C_1 \operatorname{Cos}(\mu \ln x) + C_2 \operatorname{Sin}(\mu \ln x)]x^{\lambda} \tag{9.22}$$

We shall show in the solved problems how each of these solutions is developed.

The Method of Frobenius

Note that for an equation of the form

$$(x-x_0)^2 y''(x) + a(x-x_0) y'(x) + by(x) = 0 \tag{9.23}$$

the change of variable $t = x - x_0$ reduces (9.23) to (9.17). The Euler equations (9.17) and (9.23) serve as the prototype for all second-order equations having a regular singular point. More general equations having regular singular points can be solved by the *method of Frobenius* as described in the following theorem.

Theorem 9.4 Method of Frobenius

Theorem 9.4 Suppose that $x = 0$ is a regular singular point for the equation (9.10). Then there exists at least one solution of the form,

$$y(x) = x^r \sum_{n=0}^{\infty} a_n x^n \qquad a_0 \neq 0,\ x > 0\,. \tag{9.24}$$

This series converges (at least) on the interval $0 \leq x < R$, where R is the distance from x=0 to the next singular point for (9.10).

INDICIAL EQUATION

This theorem is stated for the case in which the singular point located at x=0. If the singular point is at x_0 different from zero, then the change of variable $t=x-x_0$ reduces the problem to one with the singular point at t=0. The constant r in the solution (9.24) is obtained by solving the *indicial* equation, a quadratic equation analogous to (9.19). The coefficients a_n are obtained in the usual way from the recursion relation.

In addition to the solution of the form (9.24), there is a second solution that may be constructed by modifying the method of Frobenius. As in the case of the Euler equation, the form of the solution is determined by the character of the roots to the indicial equation. We shall consider only the case of *real* roots to the indicial equation.

Theorem 9.5

Theorem 9.5 Suppose x=0 is a regular singular point for (9.10) and that the indicial equation has real roots r_1 and $r_2 \leq r_1$. Then equation (9.10) has linearly independent solutions $y_1(x)$ and $y_2(x)$, where $y_1(x)$ is given by (9.24) and $y_2(x)$ is given by one of the following:

(a) for $r_1 - r_2$ not equal to an integer,

$$y_2(x) = x^{r_2} \sum_{n=0}^{\infty} b_n x^n$$

(b) for $r_1 = r_2$,

$$y_2(x) = y_1(x) \ln x + x^{r_1} \sum_{n=0}^{\infty} c_n x^n$$

(c) for $r_1 - r_2$ = a positive integer ,

$$y_2(x) = d_{-1} y_1(x) \ln x + x^{r_2} \sum_{n=0}^{\infty} d_n x^n$$

In each case, the series converge on the interval $0 \leq x < R$ where R is the distance from x=0 to the next singular point for (9.10). The coefficients b_n, c_n, and d_n are obtained by solving a recursion relation induced by the differential equation.

SOLVED PROBLEMS

Solutions About an Ordinary Point

PROBLEM 9.1

Solve the initial value problem

$$y''(x) - xy'(x) - y(x) = 0, \qquad y(0)=2,\ y'(0)=6 \tag{1}$$

SOLUTION 9.1

The fact that the initial conditions are imposed at x=0 dictates that we express the solution as a power series about x=0. This point is an ordinary point for equation (1) and, in fact, equation (1) has no singular points. Then Theorem 9.3 implies that we will find two power series with infinite radius of convergence defining linearly independent solutions $Y_0(x)$ and $Y_1(x)$.

We start with

$$y(x) = \sum_{n=0}^{\infty} a_n x^n \tag{2}$$

Substituting this into the differential equation (1) leads to

$$\sum_{n=2}^{\infty} n(n-1)a_n x^{n-2} - \sum_{n=1}^{\infty} n\, a_n x^n - \sum_{n=0}^{\infty} a_n x^n = 0 . \tag{3}$$

As we did in Example 9.4, we note that the following two sums produce the same series of terms

$$\sum_{n=2}^{\infty} n(n-1)a_n x^{n-2} = \sum_{n=0}^{\infty} (n+2)(n+1)\, a_{n+2} x^n . \tag{4}$$

The sum on the right can be obtained from the sum on the left by replacing the index n on the left everywhere by the index n+2. This is analogous to a translation change of variables in a definite integral.

Using the replacement (4) in (3), we find

$$2a_2 - a_0 + \sum_{n=1}^{\infty} ((n+2)(n+1)\, a_{n+2} - (n+1)a_n)\, x^n = 0 . \tag{5}$$

This equation can be satisfied for all x if and only if

$$a_2 = 1/2\, a_0 \quad \text{and} \quad a_{n+2} = \frac{1}{n+2} a_n \quad \text{for } n=1,2,... \tag{6}$$

Equation (6) is the recursion relation for the coefficients a_n in the series (2). We can calculate

$$a_4 = \frac{1}{4} a_2 = \frac{1}{4 \cdot 2} a_0 \qquad a_3 = \frac{1}{3} a_1$$

$$a_6 = \frac{1}{6} a_4 = \frac{1}{6 \cdot 4 \cdot 2} a_0 \qquad a_5 = \frac{1}{5} a_3 = \frac{1}{5 \cdot 3} a_1$$

etc.,

and then

$$\begin{aligned} y(x) &= a_0 \left[1 + \frac{1}{2} x^2 + \frac{1}{4 \cdot 2} x^4 + \frac{1}{6 \cdot 4 \cdot 2} x^6 + ...\right] \\ &+ a_1 \left[x + \frac{1}{3} x^3 + \frac{1}{5 \cdot 3} x^5 + ...\right] = a_0 Y_0(x) + a_1 Y_1(x) . \end{aligned} \tag{7}$$

Without applying any tests, we are assured by Theorem 9.3 that each of the series in (7) has radius of convergence equal to infinity. Since it is evident that

$$Y_0(0) = a_0 \qquad Y_1(0) = 0$$

$$Y_0'(0) = 0 \qquad Y_1'(0) = a_1 ,$$

the initial conditions in (1) imply that $a_0=2$ and $a_1=6$; i.e.,

$$y(x) = 2\, Y_0(x) + 6\, Y_1(x).$$

The functions Y_0 and Y_1 are linearly independent since their Wronskian at x=0 is equal to 12.

PROBLEM 9.2

Solve the initial value problem,

$$y''(x) - xy'(x) - y(x) = 0, \qquad y(1)=2, \quad y'(1)=6. \tag{1}$$

SOLUTION 9.2

In this problem, the initial conditions suggest that we express the solution as a power series expanded about the ordinary point x =1. This is one option that is available to us, but a more convenient option is to make the change of the independent variable,

$$t = x-1.$$

Then (1) becomes

$$y''(t) - (t+1)y'(t) - y(t) = 0, \qquad y(0)=2, \quad y'(0)=6. \tag{3}$$

Since t=0 is an ordinary point for (3), we are led to suppose

$$y(t) = \sum_{n=0}^{\infty} a_n t^n .$$

If we substitute this into (3) we obtain

$$\sum_{n=2}^{\infty} n(n-1)a_n t^{n-2} - (t+1)\sum_{n=1}^{\infty} na_n t^{n-1} - \sum_{n=0}^{\infty} a_n t^n = 0$$

Thus,

$$\sum_{n=2}^{\infty} n(n-1)a_n t^{n-2} - \sum_{n=1}^{\infty} a_n t^n - \sum_{n=1}^{\infty} na_n t^{n-1} - \sum_{n=0}^{\infty} a_n t^n = 0 \tag{4}$$

In order to facilitate the equating of like powers of t in these sums, we shift the index in the first and third sums appearing in (4). That is, we make the replacements

$$\sum_{n=2}^{\infty} n(n-1)a_n t^{n-2} = \sum_{n=0}^{\infty} (n+2)(n+1)a_{n+2} t^n$$

and

$$\sum_{n=1}^{\infty} na_n t^{n-1} = \sum_{n=0}^{\infty} (n+1)a_{n+1} t^n$$

in equation (4). Then (4) becomes,

$$2a_2 - a_1 - a_0 + \sum_{n=1}^{\infty} ((n+2)(n+1)a_{n+2} - (n+1)a_{n+1} - na_n - a_n)\, t^n = 0 .$$

This equation is satisfied for all values of t if and only if

$$a_2 = (a_1 + a_0)/2 \qquad a_{n+2} = \frac{1}{n+2}(a_{n+1} + a_n)\,. \tag{5}$$

Now we are able to compute from (5)

$$a_3 = \frac{1}{3}(a_2 + a_1) = \frac{1}{3}\left(\frac{1}{2}(a_1 + a_0) + a_1\right) = \frac{1}{2}a_1 + \frac{1}{6}a_0$$

$$a_4 = \frac{1}{4}(a_3 + a_2) = \frac{1}{4}a_1 + \frac{1}{6}a_0$$

$$a_5 = \frac{1}{5}(a_4 + a_3) = \frac{3}{20}a_1 + \frac{2}{30}a_0$$

etc.

Then

$$y(t) = a_0 + a_1 t + \frac{1}{2}(a_1 + a_0)t^2 + \left(\frac{1}{2}a_1 + \frac{1}{6}a_0\right)t^3 + \left(\frac{1}{4}a_1 + \frac{1}{6}a_0\right)t^4$$

$$+ \left(\frac{3}{20}a_1 + \frac{2}{30}a_0\right)t^5 + \ldots$$

$$= a_0\,[1 + 1/2\,t^2 + 1/6\,t^3 + 1/6\,t^4 + 2/30\,t^5 + \ldots\,]$$

$$+ a_1\,[1 + 1/2\,t^2 + 1/2\,t^3 + 1/4\,t^4 + 3/20\,t^5 + \ldots\,]$$

$$= a_0 Y_0(t) + a_1 Y_1(t).$$

The functions Y (t) and Y (t) satisfy

$$Y_0(0) = a_0 \qquad Y_1(0) = 0$$

$$Y'_0(0) = 0 \qquad Y'_1(0) = a_1$$

hence the initial conditions imply $a_0=2$, $a_1=6$. Then, returning to the original independent variable, $t=x-1$

$$y(x) = 2\,Y_0(x-1) + 6\,Y_1(x-1). \tag{6}$$

The functions Y_0 and Y_1 are linearly independent solutions since their Wronskian is not zero. Since Theorem 4.3 implies that the solution to the initial value problem (1) is unique, we can be assured that the solution (6) is the same solution we would have obtained had we chosen to expand y(x) about x=1 instead of changing the variable as we did.

PROBLEM 9.3

Find a power series expansion about x=0 for the general solution of the differential equation

$$(1+x^2)y''(x) + xy'(x) + 2y(x) = 0\,. \tag{1}$$

SOLUTION 9.3

Since $p(x)=1+x^2$ does not vanish at $x=0$, we see that $x=0$ is an ordinary point for the equation (1). Note that $p(x)$ does vanish at $x=\pm i$, and so the points $x=i$ and $x=-i$ are singular points for the equation (1). Since the distance from $x=0$ to $x=\pm i$ is 1, the radius of convergence of the series we are about to construct is not less than one.

If we suppose

$$y(x) = \sum_{n=0}^{\infty} a_n x^n , \tag{2}$$

then it follows from (1) that

$$\sum_{n=2}^{\infty} n(n-1)a_n x^{n-2} + \sum_{n=1}^{\infty} n(n-1)a_n x^n + \sum_{n=1}^{\infty} na_n x^n + \sum_{n=0}^{\infty} 2a_n x^n = 0 .$$

We shift the index in the first sum in this equation so that all sums are stated in terms of the same power of x. This leads to

$$\sum_{n=0}^{\infty} (n+2)(n+1)a_{n+2}x^n + \sum_{n=2}^{\infty} n(n-1)a_n x^n + \sum_{n=1}^{\infty} na_n x^n + \sum_{n=0}^{\infty} 2a_n x^n = 0 .$$

That is,

$$2a_2 + 3 \cdot 2\, a_3 x + \sum_{n=2}^{\infty} (n+2)(n+1)a_{n+2}x^n +$$

$$+ \sum_{n=2}^{\infty} n(n-1)a_n x^n + a_1 x + \sum_{n=2}^{\infty} na_n x^n + 2a_0 + 2a_1 x + \sum_{n=2}^{\infty} 2a_n x^n = 0 .$$

It follows now that this equation holds for all x near $x=0$ if and only if

$$2a_2 + 2a_0 = 0, \qquad 6a_3 + 3a_1 = 0,$$

$$(n+2)(n+1)\, a_{n+2} + (n^2+2)a_n = 0.$$

That is

$$a_{n+2} = -\frac{n^2+2}{(n+2)(n+1)}\, a_n \qquad \text{fpr } n=0,1, \ldots \tag{4}$$

and we compute

$$a_2 = -a_0 \qquad\qquad a_3 = -\frac{1}{2}\, a_1$$

$$a_4 = -\frac{6}{4 \cdot 3}\, a_2 = \frac{1}{2}\, a_0 \qquad\qquad a_5 = -\frac{11}{5 \cdot 4}\, a_3 = \frac{11}{5 \cdot 4 \cdot 2}\, a_1$$

etc.

Then

$$y(x) = a_0\,[1 - x^2 + \frac{1}{2}\,x^4 - \dots\,] + a_1\,[x - \frac{1}{2}\,x^3 + \frac{11}{5\cdot 4\cdot 2}x^5 - \dots\,]$$

is the required power series expansion for the solution to (1). Theorem 9.1 guarantees that the radius of convergence of this series is at least 1.

Legendre's Equation

PROBLEM 9.4

Find a power series expansion about x=0 for the general solution to the differential equation

$$(1-x^2)y''(x) - 2xy'(x) + N(N+1)y(x) = 0. \tag{1}$$

Here N denotes a positive integer.

SOLUTION 9.4

This equation has singular points at $x=\pm 1$, but x=0 is an ordinary point. Then the series we are about to construct will have a radius of convergence not less than one. If we suppose that

$$y(x) = \sum_{n=0}^{\infty} a_n x^n , \tag{2}$$

then substituting this into (1) gives

$$\sum_{n=2}^{\infty} n(n-1)a_n x^{n-2} - \sum_{n=2}^{\infty} n(n-1)a_n x^n - \sum_{n=1}^{\infty} 2na_n x^n + N(N+1)\sum_{n=0}^{\infty} a_n x^n = 0\,.$$

We need only shift the index in the first sum in this equation. Then the equation becomes

$$\sum_{n=0}^{\infty} (n+2)(n+1)a_{n+2}x^n - \sum_{n=2}^{\infty} n(n-1)a_n x^n - \sum_{n=1}^{\infty} 2na_n x^n + N(N+1)\sum_{n=0}^{\infty} a_n x^n = 0\,.$$

Finally, we break off *two* terms each from the first and last sums and *one* term from the third sum so that all the sums begin with n=2. Then we have

$$2a_2 + 6a_3 - 2a_1x + N(N+1)(a_0+a_1x) +$$

$$+\sum_{n=2}^{\infty} [(n+2)(n+1)a_{n+2} - n(n-1)a_n - 2na_n + N(N+1)a_n]x^n = 0\,.$$

We obtain the recursion relation

$$a_{n+2} = \frac{n(n+1) - N(N+1)}{(n+2)(n+1)}\,a_n \tag{3}$$

$$\frac{(n-N)(n+N+1)}{(n+2)(n+1)}\,a_n$$

for the coefficients in the series (2). Note that a_{n+2} vanishes so that one of the two infinite series $Y_0(x)$, $Y_1(x)$ that arises out of (3) must terminate after N terms; i.e., it is a polynomial of degree N. The equation (1) is called *Legendre's* equation and the Nth degree polynomial is called a *Legendre polynomial.* In particular, if N is even and we take $a_1 = 0$, then (1) is a polynomial of degree N containing only *even* powers of x. If N is odd and we take $a_0 = 0$, then (1) is a polynomial of degree N containing only *odd* powers of x. In either case, since the series contains only finitely many terms, there is no question of convergence. If we denote the Legendre polynomial corresponding to integer N by $P_N(x)$, then the first few Legendre polynomials are

$$P_0(x) = 1, \quad P_1(x) = x, \quad P_2(x) = (3x^2-1)/2$$

$$P_3(x) = (5x^3-3x)/2 \qquad P_4(x) = (35x^4-30x^2+3)/6 .$$

Although one of the series generated by (3) terminates at N terms, the other series does not terminate and has a radius of convergence not less than one. If N is not an integer, then *neither* of the series will terminate.

Particular Solutions

PROBLEM 9.5

Find a power series expansion about x=1 for the general solution of the inhomogeneous equation

$$y''(x) + 2(x-1)\, y'(x) + 3y(x) = 4x. \qquad (1)$$

SOLUTION 9.5

The equation (1) has no singular points, hence x=1 is an ordinary point. The inhomogeneous term 4x is a polynomial, hence it is analytic at every point, including x=1. Then the equation (1) has two linearly independent solutions that can be expanded in power series about x=1. However, it will be more convenient to change the independent variable to t=x–1 so that we can expand about t=0. Since dt/dx=1, the equation transforms to

$$y''(t) + 2ty'(t) + 3\, y(t) = 4t+4 \qquad (2)$$

Assuming

$$y(t) = \sum_{n=0}^{\infty} a_n t^n \qquad (3)$$

and proceeding in the usual way, we find

$$\sum_{n=0}^{\infty} [(n+2)(n+1)\, a_{n+2} + (2n+3)a_n]\, t^n = 4t + 4 \, ;$$

i.e.,

$$(2a_2+3a_0) + (6a_3+5a_1)t + \sum_{n=2}^{\infty} [(n+2)(n+1)\, a_{n+2} + (2n+3)a_n]\, t^n = 4t + 4 .$$

Then in order for this equation to hold for all values of t near t=0, it is necessary and sufficient that

$$2a_2 + 3a_0 - 4 = 0, \qquad 6a_3 + 5a_1 - 4 = 0, \tag{4}$$

$$a_{n+2} = -\frac{2n+3}{(n+2)(n+1)} a_n$$

These recursion relations produce the following coefficients:

$$a_2 = 2 - \frac{3}{2} a_0 \qquad a_3 = \frac{2}{3} - \frac{5}{6} a_1$$

$$a_4 = -\frac{7}{4 \cdot 3} a_2 = -\frac{7}{6} + \frac{21}{4!} a_0$$

$$a_5 = -\frac{9}{5 \cdot 4} a_3 = -\frac{3}{10} + \frac{45}{5!} a_1$$

etc.

Then

$$y(x) = a_0 Y_0(x) + a_1 Y_1(x) + [2t^2 + \frac{2}{3} t^3 - \frac{7}{6} t^4 - \frac{3}{10} t^5 + \ldots]$$

$$= a_0 Y_0(t) + a_1 Y_1(t) + Y_p(t) .$$

Finally, we make the replacement t=x–1 in order to get a solution in terms of the original variable x. Each series in the solution has infinite radius of convergence since the equation has no singular points and the forcing term is a polynomial.

PROBLEM 9.6

Find a power series expansion about x=0 for the solution of the initial value problem

$$y''(x) - xy'(x) - y(x) = 3e^x, \quad y(0) = 2, \quad y'(0)=5 \tag{1}$$

SOLUTION 9.6

This equation has no singular points so x=0 is a regular point. In addition, the inhomogeneous term $3e^x$ has a convergent power series about x=0. Then we may assume the equation (1) has a solution in the form of a power series expansion about x=0. Substituting the series into (1) leads to

$$\sum_{n=2}^{\infty} n(n-1)a_n x^{n-2} - \sum_{n=1}^{\infty} n a_n x^n - \sum_{n=0}^{\infty} a_n x^n = \sum_{n=0}^{\infty} 3 \frac{x^n}{n!} .$$

Note that we have expanded the function $3e^x$ in a power series about x=0 as well. Then, proceeding as in the previous examples to shift the index in the first sum, we obtain

$$\sum_{n=0}^{\infty} (n+2)(n+1)a_{n+2}x^n - \sum_{n=1}^{\infty} na_nx^n - \sum_{n=0}^{\infty} a_nx^n = \sum_{n=0}^{\infty} 3\frac{x^n}{n!}.$$

Then

$$2a_2 - a_0 - 3 + \sum_{n=1}^{\infty} ((n+2)(n+1)a_{n+2} - (n+1)a_n - \frac{3}{n!})x^n = 0,$$

and it follows that

$$a_2 = (a_0 + 3)/2$$

$$a_{n+2} = \frac{3}{n!}\,\frac{1}{(n+2)(n+1)} + \frac{1}{n+2}\,a_n \qquad n=1,2,...$$

We compute

$$a_3 = \frac{1}{2} + \frac{1}{3}a_1$$

$$a_4 = \frac{1}{8} + \frac{1}{4}a_2 = \frac{1}{2} + \frac{1}{8}a_0$$

$$a_5 = \frac{1}{40} + \frac{1}{5}a_3 = \frac{1}{8} + \frac{1}{5\cdot 3}a_1 \quad \text{etc.}$$

Then

$$y(x) = a_0\,[1 + \frac{1}{2}x^2 + \frac{1}{8}x^4 + ...] + a_1\,[x + \frac{2}{3}x^3 + \frac{2^3}{5!}x^5 + ...]$$

$$+ [\frac{3}{2}x^2 + \frac{1}{2}x^3 + \frac{1}{2}x^4 + \frac{1}{8}x^5 + ...]$$

$$= a_0Y_0(x) + a_1Y_1(x) + Y_p(x).$$

Since

$$Y_0(0)=1, \qquad Y_1(0)=Y_p(0)=0$$

$$Y'_1(0)=1, \qquad Y'_0(0)=Y'_p(0)=0,$$

it follows that $a_0=2$ and $a_1=5$, and the required solution is

$$y(x) = 2\,Y_0(x) + 5\,Y_1(x) + Y_p(x).$$

Each of these series has an infinite radius of convergence.

Solutions Near Singular Points

PROBLEM 9.7

Find the general solution for each of the following Euler equations:

(a) $x^2y''(x) - 4xy'(x) + 4y(x) = 0, \qquad x > 0,$

(b) $x^2y''(x) - 3xy'(x) + 4y(x) = 0, \qquad x > 0,$

(c) $x^2y''(x) + xy'(x) + 4y(x) = 0, \qquad x > 0.$

SOLUTION 9.7

In general, substituting the trial solution x^r into the equation

$$L[y(x)] = x^2y''(x) + axy'(x) + by(x) = 0, \qquad x>0 \tag{1}$$

leads to

$$L[x^r] = r(r-1)x^r + brx^r + ax^r = P(r)x^r = 0. \tag{2}$$

This equation is satisfied for $x>0$ if $P(r)= 0$; i.e., if r satisfies

$$P(r) = r^r - (1-a)r + b = 0. \tag{3}$$

The roots of this quadratic equation are

$$r_{1,2} = \frac{1-a \pm \sqrt{(1-a)^2 - 4b}}{2}. \tag{4}$$

In the case that $(1-a)^2-4b=0$, we have a double root; i.e., $P(r)=(r-r_1)^2$. Then $P(r_1)=P'(r_1)=0$. It follows that if we differentiate both sides of the equation (2) with respect to the parameter r, then

$$\partial/\partial r\, L[x^r] = L[\partial x^r/\partial r] = L[x^r \ln x] = P'(r)\, x^r + P(r)\, x^r \ln x$$

and

$$L[x^{r_1} \ln x] = P'(r_1)x + P(r_1)x \ln x = 0.$$

This indicates that if r is a repeated root then

$$y_1(x) = x^r \quad \text{and} \quad y_2(x) = x^r \ln x$$

are two linearly independent solutions of (1).

In the case of equation (a), we have

$$P(r) = r^2 - 5r + 4 = (r-1)(r-4) = 0,$$

and the general solution is then

$$y(x) = C_1x + C_2x^4.$$

In the case of equation (b),

$$P(r) = r^2 - 4r + 4 = (r-2)^2 = 0,$$

which leads to the general solution,

$$y(x) = C_1 x^2 + C_2 x^2 \ln x .$$

Finally, for equation (c) we find

$$P(r) = r^2 + 4 = (r-2i)(r+2i) = 0.$$

If we use the identity, $e^{ix} = \mathrm{Cos}x + i\mathrm{Sin}x$ to write,

$$x^{2i} = e^{i2\ln x} = \mathrm{Cos}(2\ln x) + i\mathrm{Sin}(2\ln x),$$

then the general solution for (c) can be expressed as

$$y(x) = C_1\,\mathrm{Cos}(2\ln x) + C_2\,\mathrm{Sin}(2\ln x).$$

Note that while the equations in (a), (b), and (c) are quite similar, the solutions to these equations behave quite differently from one another.

PROBLEM 9.8

Find the solution of the initial value problem

$$4x^2y''(x) + 2x(1-x)y'(x) + 6x\,y(x) = 0, \qquad y(0)= 3, \qquad \lim_{x\to 0+} \sqrt{x}\,y'(x) = 1 . \tag{1}$$

SOLUTION 9.8

Note that each of the functions

$$x\,\frac{q(x)}{p(x)} = \frac{1-x}{2} \quad \text{and} \quad x^2\,\frac{r(x)}{p(x)} = \frac{3x}{2}$$

has a finite limit as x tends to zero. Thus, x=0 is a regular singular point for the equation (1), and we may suppose that there is at least one solution of the form

$$y(x) = x^r \sum_{n=0}^{\infty} a_n x^n . \tag{2}$$

In addition to finding the coefficients a_n in the expansion, we must also find the value of the exponent r. Since x=0 is the only singular point for the equation (1), it follows from Theorem 9.4 that the series (2), once it has been found, will converge for all *positive* values of x. Substituting (2) into (1) leads to

$$L[y(x)] = [4r(r-1)a_0x^r + 4(r+1)ra_1x^{r+1} + 4(r+2)(r+1)a_2x^{r+2} + \dots] +$$
$$+ [2ra_0x^r + 2(r+1)a_1x^{r+1} + 2(r+2)a_2x^{r+2} + \dots] -$$
$$- [2ra_0x^{r+1} + 2(r+1)a_1x^{r+2} + \dots] +$$
$$+ [6a_0x^{r+1} + 6a_1x^{r+2} + \dots] = 0$$

Thus,

$$L[y(x)] = (4r^2-2r)a_0x^r + [(4r+2)(r+1)a_1 + (6-2r)a_0]x^{r+1} +$$
$$+ [(4r+6)(r+2)a_2 + (6-2(r+1))a_1]x^{r+2} + \dots = 0.$$

By equating the coefficients of successive powers of x to zero, we obtain the following sequence of equations for the unknown ingredients:

$$2r(2r-1)a_0 = 0$$
$$2(r+1)(2r+1)a_1 + 2(3-r)a_0 = 0 \tag{3}$$
$$2(r+2)(2r+3)a_2 + 2(2-r)a_1 = 0$$

etc.

We want the coefficient a_0 to remain arbitrary if possible, hence the first of these equations is satisfied if

$$2r(2r-1) = 0. \tag{4}$$

This is the so called *indicial equation* which determines the values of the exponent r in (2). The remaining equations in (3) can be stated in the following compressed form:

$$2(r+k)(2r+2k-1)a_k + (6-2(r+k-1))a_{k-1} = 0;$$

Thus,

$$a_k = -\frac{6-2(r+k-1)}{2(r+k)(2r+2k-1)}\, a_{k-1} \qquad \text{for } k=1,2,\dots \tag{5}$$

The indicial equation (4) has solutions $r=0, 1/2$. Since the roots do not differ by an integer, it follows from Theorems 9.4 and 9.5 that this equation has two solutions of the Frobenius form, (9.24). For $r=0$, the recursion relation (5) reduces to

$$a_k = -\frac{4-k}{k(2k-1)}\, a_{k-1} \qquad \text{for } k=1,2,\dots \tag{6}$$

That is,

$$a_1 = -3a_0$$
$$a_2 = -a_1/3 = a_0$$
$$a_3 = -a_2/(3\cdot 5) = -a_0/15$$
$$a_4 = 0 \quad \text{and} \quad a_k = 0 \text{ for } k=4,5,\dots$$

Then, setting a_0 equal to 1,

$$y_1(x) = 1-3x+x^2-x^3/15\,.$$

For r=1/2, the recursion relation reduces to

$$a_k = -\frac{7-2k}{2k(2k+1)}\, a_{k-1}\,. \tag{7}$$

This leads to

$$y_2(x) = \sqrt{x}\,(1 - \frac{5}{3!}\,x + \frac{3\cdot5}{5!}\,x^2 - \frac{3\cdot5}{7!}\,x^3 - \frac{3\cdot5}{9!}\,x^4 - \frac{9\cdot5}{11!}\,x^5 - \ldots).$$

Then the general solution for the equation is given by

$$y(x) = C_1y_1(x) + C_2y_2(x).$$

Since

$$y_1(0)=1\,, \qquad y_2(0)=0$$

$$y'_1(0)=-3, \qquad \lim_{x\to0+} \sqrt{x}\,y_2(x) = 1/2\,,$$

it follows that the initial conditions are satisfied if $C_1=3$ and $C_2=2$.

PROBLEM 9.9

Find the general solution of the equation

$$4xy''(x)+(x+1)y'(x)+2y(x)= 0 \tag{1}$$

that is valid in a neighborhood of x=0.

SOLUTION 9.9

In order for the solution we construct to be valid in a neighborhood of x=0, we will expand in a power series about zero. Since

$$A(x) = \frac{x+1}{4x} \qquad \text{and} \qquad B(x) = \frac{2}{4x}\,,$$

it is evident that x=0 is a regular singular point for the equation (1) hence there is at least one solution of the form (9.24). In addition, since x=0 is the *only* singular point, this series will converge on the interval $0 < x < \infty$.

When we substitute the series (9.24) into the equation (1), we obtain

$$\sum_{n=0}^{\infty} 4(n+r)(n+r-1)a_nx^{n+r-1} + \sum_{n=0}^{\infty}(n+r)a_nx^{n+r-1} + \sum_{n=0}^{\infty}(n+r)a_nx^{n+r} + \sum_{n=0}^{\infty}2a_nx^{n+r} = 0,$$

which implies that

$$[4(r-1)+1]\,ra_0 = 0$$

$$(4r+1)(r+1)a_1 + (r+2)a_0 = 0$$

$$[4(r+1)+1](r+2)a_2 + [(r+1)+2]a_1 = 0,$$

etc.

The first of these equations is the indicial equation. It is satisfied for arbitrary a_0 if r=0 or if r= 3/4. Since r_1-r_2 does not equal an integer, there are two solutions of the form (9.24). The recursion relation for the coefficients in the series is found to be

$$a_{k+1} = -\frac{k+r+2}{4(k+r)+1} a_k \qquad \text{for } k=0,1,\ldots$$

For r=0 this produces

$$a_1 = -2a_0$$

$$a_2 = -\frac{3}{5} a_1 = \frac{6}{5} a_0$$

$$a_3 = -\frac{4}{9} a_2 = -\frac{6 \cdot 4}{9 \cdot 5} a_0$$

$$a_4 = -\frac{5}{13} a_3 = \frac{6 \cdot 5 \cdot 4}{13 \cdot 9 \cdot 5} a_0$$

etc.

Then

$$y_1(x) = 1 - 2x + \frac{6}{5} x^2 - \frac{6 \cdot 4}{9 \cdot 5} x^3 + \frac{6 \cdot 5 \cdot 4}{13 \cdot 9 \cdot 5} x^4 + \ldots .$$

For r=3/4, the recursion relation produces a sequence of coefficients leading to the solution,

$$y_2(x) = x^{3/4} \left(1 - \frac{11}{16} x + \frac{15 \cdot 11}{2 \cdot 16 \cdot 16} x^2 - \frac{19 \cdot 15 \cdot 11}{3! \cdot 16 \cdot 16 \cdot 16} x^3 + \ldots\right).$$

Then the general solution can be expressed as, $y(x) = C_1y_1(x) + C_2y_2(x)$.

Bessel's Equation

PROBLEM 9.10

Find a general solution, valid near x=0, for Bessel's equation

$$x^2y''(x) + xy'(x) + (x^2-\alpha^2)y(x) = 0 \qquad \alpha \neq \text{integer.} \qquad (1)$$

SOLUTION 9.10

Bessel's equation arises in numerous applications in applied mathematics, particularly in connection with solving partial differential equations that involve the Laplace operator expressed in cylindrical coordinates.

Since $xA(x)=1$ and $x^2B(x)=(x^2-\alpha^2)$, each has a finite limit as x approaches zero, and x=0 is a regular singular point for Bessel's equation (1). Substituting a series of the form (9.24) into (1) leads to

$$(r^2-\alpha^2)a_0x^r + [(r+1)^2-\alpha^2]a_1x^{r+1} + \sum_{n=2}^{\infty} (((n+r)^2-\alpha^2)a_n+a_{n-2})x^{n+r} = 0.$$

It follows that

$$(r^2-\alpha^2)a_0 = 0,$$
$$((r+1)^2 - \alpha^2)a_1 = 0$$
$$((n+r)^2 - \alpha^2)a_n + a_{n-2} = 0 \qquad \text{for } n=2,3, \ldots$$

The first two equations are satisfied if

$$r=\alpha, -\alpha, \quad a_0=\text{arbitrary}, \quad a_1=0, \tag{2}$$

and then the remaining coefficients are determined from the recursion relation

$$a_n = -\frac{1}{(n+r)^2 - \alpha^2}\, a_{n-2} \tag{3}$$

For r=a, (3) reduces to

$$a_n = -\frac{1}{(n+2\alpha)\, n}\, a_{n-2} \qquad n=2,3,\ldots$$

which produces the following series:

$$y_1(x) = a_0x^\alpha \left[1-(\frac{x}{2})^2 \frac{1}{1+\alpha} +(\frac{x}{2})^4 \frac{1}{2!} \frac{1}{(1+a)(2+a)} - -(\frac{x}{2})^6 \frac{1}{3!} \frac{1}{(1+\alpha)(2+\alpha)(3+\alpha)} + \ldots\right]. \tag{4}$$

The solution (4) can be expressed more conveniently in terms of the *Gamma function*

$$\Gamma(x) = \int_0^\infty e^{-t}\, t^{x-1}\, dt \qquad x>0\, .$$

It can be shown that $\Gamma(x)$ satisfies

$$\Gamma(x+1) = x\Gamma(x) \quad \text{for } x>0 \quad \text{and } \Gamma(1)=1\, .$$

In particular, when x=n, an integer, $\Gamma(n+1)=n!$. Our interest in the gamma function lies in the fact that for $\alpha>0$,

$$\Gamma(\alpha+n+1) = \Gamma(\alpha+1)(1+\alpha)(2+\alpha) \ldots (n+\alpha). \tag{5}$$

If we choose the arbitrary constant a_0 in (4) equal to $1/[2^\alpha\Gamma(\alpha+1)]$, then (4) may be written as

$$y_1(x) = \sum_{n=0}^{\infty} (-1)^n \left(\frac{x}{2}\right)^{2n+\alpha} \frac{1}{n!\Gamma(\alpha+n+1)} = J_\alpha(x) \tag{6}$$

The series that appears in (6) is called the *Bessel function of the first kind of order* α. It is usually denoted by the symbol $J_\alpha(x)$.

Since the roots of the indicial equation, α and $-\alpha$, differ by an amount 2α that is not equal to an integer, Theorem 9.5 ensures that a second solution of the form (9.24) may be obtained from the smaller root; i.e., for $r=-\alpha$. In this case, the recursion relation (3) reduces to

$$a_n = -\frac{1}{(n-2\alpha)n} a_{n-2} \qquad n=2,3,...$$

It is not difficult to show that this leads eventually to the solution

$$y_2(x) = J_{-\alpha}(x),$$

where $J_{-\alpha}(x)$ is obtained by formally replacing α with $-\alpha$ in (6). Then the general solution of the Bessel equation (1), in the case that α is not an integer, is

$$y(x) = C_1J_\alpha(x) + C_2J_{-\alpha}(x).$$

Note that $J_\alpha(0)=0$, while $J_{-\alpha}(x)$ grows without bound as x tends to zero. Thus, the boundary condition that the solution y(x) remain bounded at x=0 would require that $C_2=0$. Since x=0 is the only singular point for Bessel's equation, the series for $J_\alpha(x)$ and $J_{-\alpha}(x)$ converge for all positive values of x.

If α is equal to an integer m then $J_{-m}(x)=(-1)^mJ_m(x)$. Then when α is an integer, there is only one independent solution of the form (6). For an indication of how a second independent solution may be obtained, see Problem 9.12.

PROBLEM 9.11

Find a series expansion about x=0 for the solution of

$$2xy''(x) + 2y'(x) - xy(x) = 0 \tag{1}$$

SOLUTION 9.11

It is easily checked that x=0 is a regular singular point for this equation. Then there is at least one solution of (1) that can be expanded in a series of the form (9.24). If we substitute (9.24) into (1), we obtain

$$2r^2a_0x^{r-1} + 2(r+1)^2a_1x^r + \sum_{n=0}^{\infty} [2(r+2)^2a_{n+2} - a_n]x^{r+n+1} = 0,$$

and this in turn implies

$$2r^2a_0 = 0$$

$$2(r+1)^2a_1 = 0$$

$$2(r+2)^2\,a_{n+2} - a_n = 0 \qquad \text{for } n=0,1,...$$

Then

$$r^2=0, \quad a_0=\text{arbitrary}, \quad \text{and} \quad a_1=0,$$

$$a_{n+2} = a_n/(2(r+2)^2) \qquad \text{for } n=0,1,... \tag{2}$$

Since the indicial equation has the repeated root r=0, there is only one solution of the form (9.24). It is not difficult to apply the recursion relation (2) with r=0 to obtain

$$y_1(x) = a_0\,(1 + \frac{x^2}{2^3} + \frac{x^4}{2^6} + ... \tag{3}$$

In order to obtain a second linearly independent solution, we use Problem 9.7 as our guide, focusing in particular on part (b) of that problem. There we saw that in the case of a repeated root for the indicial equation, not only was $Y(x{:}r)=x^r$ a solution of the differential equation, the derivative with respect to r, $\partial_r Y(x{:}r),r$ was also a solution. Therefore, we use the recursion relation (2), with r left unspecified, to compute

$$a_2 = a_0/(2(r+2)^2)$$

$$a_4 = a_0/(2^2(r+2)^4)$$

$$a_6 = a_0/(2^4(r+2)^6)$$

etc.

Then

$$Y(x{:}r) = a_0(x^r + \frac{x^{2+r}}{2(r+2)^2} + \frac{x^{4+r}}{2^2(r+2)^4} + ...)$$

$$= a_0 \sum^{\infty} (\frac{x^2}{2})^n \frac{x^r}{(r+2^{2n}}$$

and

$$\partial_r\, Y(x{:}r) = a_0 \sum^{\infty} (\frac{x^2}{2})^n[\frac{x^r \ln x}{(r+2)^{2n}} - \frac{2nx^r}{(r+2)^{2n+1}}]$$

Now setting r=0, we obtain

$$\partial_r\, Y(x{:}0) = y_1(x)\ln x - 2a_0 \sum_{n=0}^{\infty} \frac{n}{2^n}\,(x/2)^{2n}$$

Thus,

$$y_2(x) = y_1(x)\ln x - \sum_{n=0}^{\infty} \frac{n}{2^n} (x/2)^{2n}$$

is a second independent solution of (1). Note that it is of the form indicated in part (b) of Theorem 9.5. The general solution of (1) is

$$y(x) = C_1y_1(x) + C_2y_2(x).$$

PROBLEM 9.12

Find a power series expansion about x=0 for the general solution of the Bessel equation of order zero,

$$x^2y''(x) + xy'(x) + x^2y(x) = 0 \qquad (1)$$

Note that this is the equation from Problem 9.10 in the case $\alpha=0$.

SOLUTION 9.12

The equation (1) is a special case of the equation solved in Problem 9.10. Then x=0 is a regular singular point for (1), as it was for the equation from Problem 9.10, and it follows from Theorem 9.4 that there is at least one solution of the form (9.24). However, in this case the roots of the indicial equation are the repeated values r= 0,0. The recursion relation discovered in Problem 9.10 reduces to

$$a_n = \frac{-1}{(n+r)^2}\, a_{n-2} \qquad n = 2,3,... \qquad (2)$$

for $\alpha=0$. Then, setting r=0 leads to the solution

$$y_1(x) = a_0\left(1 - \frac{x^2}{2^2} + \frac{x^4}{2^2 \cdot 4^2} - \frac{x^6}{2^2 \cdot 4^2 \cdot 6^2} + ...\right) (3)$$

Note that the choice $a_0 = 1$ leads to

$$y_1(x) = \sum^{\infty} (-1)^n \frac{(x/2)^{2n}}{(n!)^2} = J_0(x)$$

This is the single solution of Frobenius form that is guaranteed by Theorem 9.4. To find a second independent solution, we proceed as in the previous problem, using (2) with r left unspecified, to write

$$Y(x:r) = a_0 \left[x^r - \frac{x^{2+r}}{(2+r)^2} + \frac{x^{4+r}}{(2+r)^2(4+r)^2} - \frac{x^{6+r}}{(2+r)^2(4+r)^2(6+r)^2} + ...\right]$$

We now differentiate this expression with respect to r to obtain

$$\partial_r Y(x{:}r) = a_0 x^r \ln x \left[1 - \frac{x^2}{(2+r)^2} + \frac{x^4}{(2+r)^2(4+r)^2} - \dots\right]$$

$$- 2\, a_0 x^r \left[0 - \frac{x^2}{(2+r)^3} + \frac{x^4}{(2+r)^2(4+r)^2}\left(\frac{1}{2+r} + \frac{1}{4+r}\right)\right.$$

$$\left. - \frac{x^6}{(2+r)^2(4+r)^2(6+r)^2}\left(\frac{1}{2+r} + \frac{1}{4+r} + \frac{1}{6+r}\right) + \dots\right]$$

The technique of *logarithmic differentiation* is helpful in computing this derivative. Now setting r=0, this expression reduces to

$$\partial_r Y(x{:}0) = a_0 \ln x\, y_1(x)$$

$$-2a_0 \left[- \frac{x^2}{2^3} + \frac{x^4}{2^2\, 4^2}\left(\frac{1}{2} + \frac{1}{4}\right) - \frac{x^6}{2^2\, 4^2\, 6^2}\left(\frac{1}{2} + \frac{1}{4} + \frac{1}{6}\right) + \dots\right]$$

Thus,

$$\partial_r Y(x{:}0) = a_0 \ln x\, J_0(x) + a_0\, U(x).$$

Then,

$$y_2(x) = \ln x\, J_0(x) + U(x)$$

is a second independent solution for (1) of the form indicated in part (b) of Theorem 9.5. The general solution for (1) is equal to

$$y(x) = C_1 J_0(x) + C_2 y_2(x).$$

Note that the expression

$$Y_0(x) = 2/\pi[J_0(x)(\ln(x/2)+\gamma) + U(x)]$$

is also a solution of (1) and can be chosen in place of $y_2(x)$ as the second independent solution of the Bessel equation of order zero. The function $Y_0(x)$ is referred to as the *Bessel function of the second kind of order zero.* The constant γ that appears in the expression for $Y_0(x)$ is called Euler's constant and has the value γ=0.5772.

PROBLEM 9.13

Find a power series expansion about x=0 for the general solution of

$$2x^2y''(x) - 6x(1-x)y'(x) + 6xy(x) = 0 \tag{1}$$

SOLUTION 9.13

We have no difficulty in determining that x=0 is a regular singular point for this equation and that substituting the series (9.24) into (1) leads to

$$(2r^2-8r)a_0x^r + 2\sum_{n=1}^{\infty}[(r+n)((r+n-4)a_n + 3a_{n-1})]x^{n+r} = 0.$$

Then

$$2r(r-4)a_0 = 0 \tag{2}$$

is the indicial equation, and

$$(r+n-4)a_n + 3a_{n-1} = 0 \qquad \text{for } n=1,2,... \tag{3}$$

is the recursion relation for the coefficients. It is evident from (2) that the indicial roots $r=0$, and $r=4$ differ by an integer, and in this case Theorem 9.5 indicates that the larger root leads to the solution of Frobenius form. Using the value $r=4$ in (3), we find

$$y_1(x) = a_0x^4(1 - 3x + \frac{3^2}{2!}x^2 - \frac{3^3}{3!}x^3 + \frac{3^4}{4!}x^4 - ...)$$

$$= a_0x^4 \sum_{n=1}^{\infty}(-1)^n(3x)^n/n!. \tag{4}$$

In order to find a second solution for (1), note that if we substitute the other root of the indicial equation, $r=0$, into (3), we obtain

$$a_0=a_1=a_2=a_3=0$$

and

$$a_n = (-1)^{n-4}\frac{3^{n-4}}{(n-4)!}a_4 \qquad \text{for } n=5,6,...$$

This leads to the solution given in (4). In order to obtain a second solution, we use (3), with r left unspecified for the moment, and generate

$$Y(x{:}r) = a_0x^r\,[1 - \frac{3x}{r-3} + \frac{(3x)^2}{(r-3)(r-2)} - \frac{(3x)^3}{(r-3)(r-2)(r-1)} + \frac{(3x)^4}{(r-3)(r-2)(r-1)r} - ...]. \tag{5}$$

Substituting this expression into (1), we find

$$L[Y(x{:}r)] = 2a_0r(r-4)x^r + 0. \tag{6}$$

Here the recursion relation (3) implies the coefficient of x^{n+r} vanishes for $n=1,2,...$ and the remaining term on the right side of (6) is seen to vanish for $r=0$ and $r=4$. However, we have already seen that setting $r=4$ and $r=0$ in (3) does not lead to independent solutions for (1). In addition, some of the terms in the series (5) for $Y(x{:}r)$ become undefined when r is set equal to zero. We can overcome this difficulty by recalling that a_0 is arbitrary and, in fact, can depend on r. In particular, we can suppose that $a_0=b_0r$ for an arbitrary constant b_0. This produces the result

$$Y(x:r) = b_0x^r \left[r - \frac{3rx}{r-3} + \frac{(3x)^2r}{(r-3)(r-2)} - \frac{(3x)^3r}{(r-3)(r-2)(r-1)} + \frac{(3x)^4}{(r-3)(r-2)(r-1)} - \frac{(3x)^5}{(r-3)(r-2)(r-1)(r+1)} + \ldots\right]$$

$$= b_0x^r\, U(x:r). \tag{7}$$

Then

$$L[Y(x:r)] = 2b_0r^2(r-4)x^r$$

and

$$L[\partial_rY(x:r)] = 2b_0[(3r^2-8r)x^r + r^2(r-4)x^r \ln x]. \tag{8}$$

Since the right side of (8) vanishes when r=0, it follows that $\partial_rY(x:r)$, evaluated at r=0, is a solution for (1). Differentiating (7) with respect to the parameter r leads to

$$\partial_rY(x:r) = b_0x^r \ln x\, U(x:r) + b_0x^r\, \partial_rU(x:r).$$

It is straightforward but tedious to show that

$$U(x:0)=y_1(x)$$

$$\partial_rU(x:0) = 1 + x + \frac{1}{3!}\left((3x)^2 + (3x)^3 - (3x)^4\left[1 + \frac{1}{2} + \frac{1}{3}\right]\ldots\right)$$

Then we have constructed a solution for (1) that is of the form

$$y_2(x) = b_0\left[\ln x\, y_1(x) + \sum_{n=0}^{\infty} d_nx^n\right],$$

as predicted by part (c) of Theorem 9.5.

This chapter is devoted to solving equations having variable coefficients. In particular, we have considered problems of the form

$$p(x)y''(x) + q(x)y'(x) + r(x)y(x) = 0, \qquad (9.25)$$

where p(x), q(x), r(x) are polynomials having no factor in common. A point x_0 *where* $p(x_0)=0$ *is called a* singular point *for the differential equation (9.25); otherwise* x_0 *is an* ordinary point. *If* $x_0=0$ *is an ordinary point for (9.25), then the differential equation has two independent solutions of the form*

$$y(x) = \sum_{n=0}^{\infty} a_n x^n, \qquad (9.26)$$

each converging on the interval $|x|<R$ *where R denotes the distance to the nearest point in the complex plane that is a singular point for (9.25). A power series solution can be expanded about a point other than the origin, but it is generally easier to introduce the change of variable* $t=x-x_0$ *in the differential equation and then expand about* $t=0$*. The coefficients* a_n *in the series (9.26) can be found by substituting the series into the equation (9.25) and collecting coefficients of like powers of x. This leads to a* recursion relation *from which the coefficients may then be determined.*

If $x_0=0$ *is a singular point, then at least one solution of the* Frobenius form

$$y(x) = x^r \sum_{n=0}^{\infty} a_n x^n, \qquad (9.27)$$

exists, provided that $x_0=0$ *is a* regular singular point; *that is, if*

$$xA(x)=xq(x)/p(x) \quad \text{and} \quad x^2r(x)/p(x),$$

each has a finite limit as x tends to $x_0=0$*. If* x_0 *is not a regular singular point, then it is said to be an* irregular singular point. *In this case there is no general method for finding a solution for (9.25) in a neighborhood of* x_0*.*

In the case that $x_0=0$ *is a regular singular point, we substitute (9.27) into the equation (9.25) and collect coefficients of like powers of x. This leads to an* indicial equation *for determining the parameter r and to a recursion relation for the coefficients* a_n*. The character of the solution is determined by the roots* r_1,r_2 *of the indicial equation. If the difference* r_1-r_2 *is not an integer, then there are two independent solutions of the Frobenius form (9.26). If* r_1-r_2 *equals an integer or zero, then the second solution is not of the Frobenius form but involves a logarithmic term as well. A power series expansion about a regular singular point other than* $x_0=0$ *is most easily accomplished by the shift of the independent variable mentioned above.*

SUPPLEMENTARY PROBLEMS

Find a power series expansion about x=0 for the general solution or the indicated particular solution for each of the following problems:

9.1 $(1+x^2)y''(x) + 2xy'(x) - 2y(x) = 0$

9.2 $y''(x) + xy'(x) - 2y(x) = 0, \quad y(0)=0 \quad y'(0)=1$

9.3 $y''(x) - x^2y(x) = \text{Sin}x \qquad y(0)=y'(0)=0$

9.4 $x^2y''(x) + (x^2+x/2)y'(x) + xy(x) = 0$

9.5 $2x^2y''(x) + (x^2-x)y'(x) + y(x) = 0$

9.6 $3x^2y''(x) + 5xy'(x) + 3xy(x) = 0$

9.7 $x^2y''(x) + 3xy'(x) + (1+x)y(x) = 0$

9.8 $y''(x) + (1+x)y(x) = 0$

9.9 $8x^2y''(x) + 10xy'(x) + (x-1)y(x) = 0$

9.10 $x(1-x)y''(x) + 2y'(x) + 2y(x) = 0$

ANSWERS TO SUPPLEMENTARY PROBLEMS

9.1 $y(x) = C_1x + C_2 \sum_{k=0}^{\infty} (-1)^k x^{2k-1}/(2k-1)$

9.2 $y(x) = x + \frac{x^3}{3!} - \frac{x^5}{5!} + \frac{3x^7}{7!} - \frac{3 \cdot 5\, x^9}{9!} + \frac{3 \cdot 5 \cdot 7\, x^9}{11!} - \ldots$

9.3 $y(x) = \sum_{k=1}^{\infty} (-1)^k x^{2k+1}/(2k+1)!$

9.4 $y(x) = \sqrt{x}\, C_1 \sum^{\infty} (-1)^k x^k/k! + C_2 \sum^{\infty} \frac{(-1)^k x^k}{(k-1/2)(k-3/2)\ldots 3/2 \cdot 1/2}$

9.5 $y(x) = C_1 x \sum^{\infty} \frac{(-1)^k\ x^k}{1 \cdot 3 \cdot 5 \cdot \ldots (2k+1)} + C_2\sqrt{x} \sum^{\infty} \frac{(-1)^k\ x^k}{2\ k!}$

9.6 $y(x) = C_1 \sum^{\infty} \frac{(-1)^k\ 3x^k}{k! 5 \cdot 8 \cdot 11 \cdot \ldots (3k+2)} + C_2 |x|^{-2/3} \sum^{\infty} \frac{(-1)^k\ 3x^k}{k! 4 \cdot 7 \cdot \ldots (3k-2)}$

9.7 $y_1(x) = |x|^{-1} \sum^{\infty} \frac{(-1)^k\, x^k}{k!\ k!}$

$y_2(x) = \ln x\, y_1(x) - 2|x|^{-1} \sum^{\infty} \frac{(-1)^k\, x^k}{k!\ k!} \left(1 + \frac{1}{2} + \frac{1}{3} + \ldots + \frac{1}{k}\right)$

9.8 $y_1(x) = 1 - \frac{x^2}{2!} - \frac{x^3}{3!} + \frac{x^4}{4!} + \frac{4x^5}{5!} + \ldots$

$y_2(x) = x - \frac{x^3}{3!} - \frac{2x^4}{4!} + \frac{x^5}{5!} + \ldots$

9.9 $y_1(x)=x^{1/4}(1-\frac{x}{14}+\frac{x^2}{616}+\ldots)$ $y_2(x)=x^{-1/2}(1-\frac{x}{2}+\frac{x^2}{40}+\ldots)$

9.10 $y_1(x) = 1/x$ $y_2(x) = 1-x+x^2/3$.

10

Numerical Methods

In previous chapters we have discussed various means of constructing solutions for differential equations and for systems of differential equations. In Chapter 2 in particular we considered problems of the form

$$dy/dx = F(x,y(x)) \qquad a<x<b,$$
$$y(a) = A, \qquad (10.1)$$

and we saw that while a solution may exist, it is not always possible to find the solution nor to express it in terms of elementary functions. In such cases we may wish to consider numerical methods as an alternative.

TERMINOLOGY

Partitions and Approximate Solutions

A function $y=y(x)$ that satisfies the conditions of (10.1) is said to be an *exact solution* of the problem. We shall refer to a set of points $\{x_0,x_1,...,x_N\}$ as a *partition* of the interval (a,b) if

$$a=x_0 < x_1 < ... < x_N = b.$$

Finally, the values $\{y_0,y_1,...,y_N\}$ form an *approximate solution* for (10.1) on the partition $\{x_0,x_1,...,x_N\}$ if

$y_0=A$ and, for each n, $|y(x_n)-y_n|$ tends to zero as h tends to zero, where

$$h = \max_{1<n<N} |x_n - x_{n-1}|$$

In this chapter we shall always use a *uniform* partition, with $h=(b-a)/N$ and

$$x_n = a + nh \qquad n=0,1, \dots N.$$

We shall describe several algorithms for generating the numbers $\{y_0, y_1, \dots, y_N\}$. First we define various types of error.

Types of Error

LOCAL TRUNCATION ERROR

The *local truncation error* is defined as the difference $E_n=|y_{n+1}-y(x_{n+1})|$, assuming that $y_n=y(x_n)$. This is the error made at the nth step, assuming the solution is correct up to that point.

ACCUMULATED ERROR

The *accumulated error* is defined as the difference $e_n=|y_n-y(x_n)|$, assuming that $y_0=y(x_0)$. This is the error accumulated up to the nth step, assuming only that the solution starts with the correct initial value.

Figure 10.1 illustrates the distinction between E_n and e_n.

ALGORITHMS FOR CONSTRUCTING APPROXIMATE SOLUTIONS

Single Step Methods

Taylor's theorem implies that for y(x) sufficiently smooth,

$$y(x_{n+1}) = y(x_n) + y'(x_n)h + y''(x_n)h^2/2! + y^{(3)}(x_n)h^3/3! + \dots \qquad (10.2)$$

Then

$$y(x_{n+1}) \sim y(x_n) + hy'(x_n),$$

and since (10.1) implies $y'(x_n)=F(x_n,y_n)$, we can approximate the initial value problem by the following algorithm known as Euler's Method.

EULER'S METHOD

$$\begin{aligned} y_{n+1} &= y_n + hF(x_n,y_n) \quad \text{for} \quad n=0, 1, \dots, N-1 \\ y_0 &= A \end{aligned} \qquad (10.3)$$

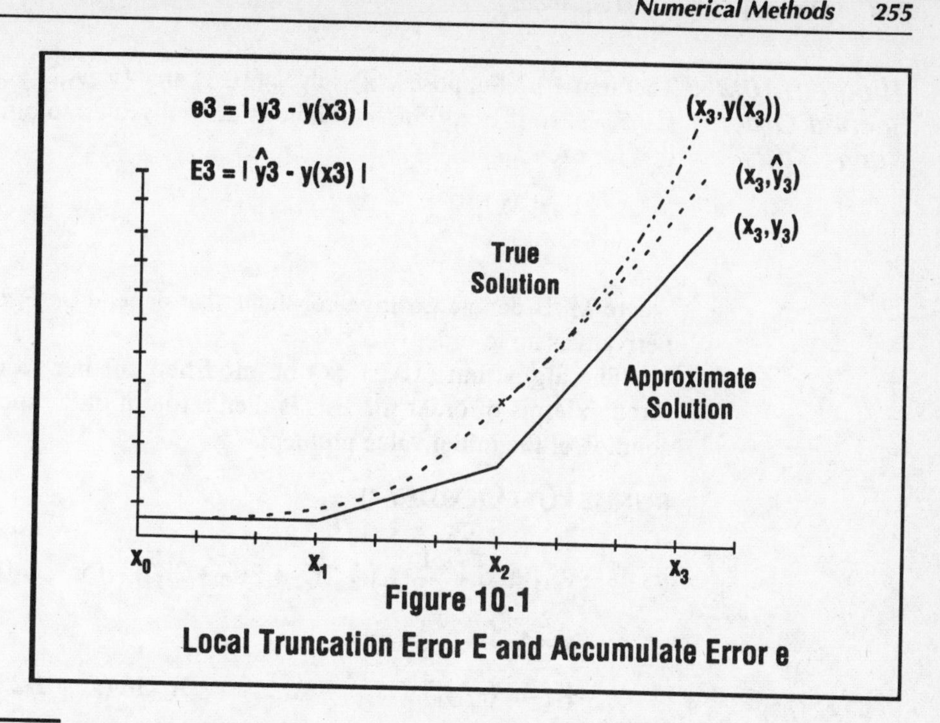

Figure 10.1
Local Truncation Error E and Accumulate Error e

Theorem 10.1 Euler's Method

Theorem 10.1 Suppose y(x) solves (10.1) and $\{y_0, y_1, \ldots, y_N\}$ is generated by (10.3). Then if F(x,y) has continuous derivatives of order one, it follows that for each n,

$$E_n \leq M\, h^2/2$$

$$e_n \leq \frac{Mh}{2B}(e^{BL}-1) \tag{10.4}$$

where L=b–a, and M, B denote positive constants that depend on F(x,y) and its partial derivatives of order one.

It is evident from (10.2) and (10.3) that Euler's method is an approximation for the solution of the initial value problem that is correct through terms of order h. The following algorithm is correct through terms of order h^2. This leads to an accumulated error that is proportional to h^2. We say that the algorithm is a method of *order two,* while Euler's method is a *first-order* method.

RUNGE KUTTA (ORDER 2)

$$y_{n+1} = y_n + \frac{1}{2}(K_1 + K_2) \quad \text{for} \quad n=0, 1, \ldots, N-1$$

$$y_0 = A \tag{10.5}$$

$$K_1 = hF(x_n, y_n), \qquad K_2 = hF(x_n+h, y_n+K_1)$$

Theorem 10.2 Second Order Runge Kutta

Theorem 10.2 Suppose y(x) solves (10.1) and $\{y_0, y_1, \ldots, y_N\}$ is generated by (10.5). Then if F(x,y) has continuous derivatives up to order two, it follows that for each n,

$$E_n \le M h^3 \qquad e_n \le Bh^2, \tag{10.6}$$

where M, B denote positive constants that depend on F(x,y) and its partial derivatives up to order two.

The algorithm (10.5) can be modified further to make it accurate through terms of order h^4. This is then a fourth-order approximation to the solution of the initial value problem.

RUNGE KUTTA (ORDER 4)

$$\begin{aligned} &y_{n+1} = y_n + \frac{1}{6}(K_1 + 2K_2 + 2K_3 + 2K_4) \quad \text{for} \quad n=0, 1, \ldots, N-1 \\ &y_0 = A \\ &K_1 = hF(x_n, y_n), \qquad K_2 = hF(x_n + 1/2\, h, y_n + 1/2\, K_1), \\ &K_3 = hF(x_n + 1/2\, h, y_n + 1/2\, K_2), \quad K_4 = hF(x_n + h, y_n + K_3) \end{aligned} \tag{10.7}$$

Theorem 10.3 Fourth Order Runge Kutta

Theorem 10.3 Suppose y(x) solves (10.1) and $\{y_0, y_1, \ldots, y_N\}$ is generated by (10.7). Then if F(x,y) has continuous derivatives up to order four, it follows that for each n,

$$E_n \le M h^5 \qquad e_n \le Bh^4 \tag{10.8}$$

where M, B denote positive constants that depend on F(x,y) and its partial derivatives up to order four.

Euler's method and each of the Runge-Kutta procedures is an example of a *single step method,* which refers to the fact that the solution value at each step is computed from the solution value at the previous step only. A solution method that computes y_n in terms of *more* than one previous value is called a *multistep method.*

Multi-Step Methods

If y(x) solves (10.1), then it follows by integration that

$$y(x_{n+1}) = y(x_n) + \int_{x_n}^{x_{n+1}} F(s,y(s))\, ds \tag{10.9}$$

The single-step algorithms we have considered are based on the following approximations for the integral in (10.9):

(a) Euler's Method $\quad E \sim Ch^2$

$$\int_{x_n}^{x_{n+1}} F(s,y(s))\,ds \sim h\,F(x_n,y_n) = hF_n$$

(b) Runge Kutta (Order 2) $\quad E \sim Ch^3$

$$\int_{x_n}^{x_{n+1}} F(s,y(s))\,ds \sim h\,[F_n + F(x_{n+1},y_n + hF_n)]/2$$

(c) Runge Kutta (Order 4) $\quad E \sim Ch^5$

$$\int_{x_n}^{x_{n+1}} F(s,y(s))\,ds \sim \frac{1}{6}\,h[F_n + 2\,F(x_n + 1/2\,h,y_n + 1/2\,K_1)$$

$$+ 2F(x_n + 1/2\,h,y_n + 1/2\,K_2) + F(x_{n+1},y_n + K_3)]$$

Now suppose that in addition to knowing the initial condition (x_0,y_0), we are given solution points (x_1,y_1) and (x_2,y_2). Then we can evaluate

$$F_n = F(x_n,y(x_n)) \qquad n=0,1,2$$

and generate the unique polynomial of degree two passing through the three solution points:

$$P(x) = [F_0(x-x_1)(x-x_2) + F_1(x-x_0)(x-x_2) + F_2(x-x_0)(x-x_1)]/(2h^2)\,.$$

Replacing $F(s,y(s))$ with $P(s)$ in (10.9) and integrating leads to

$$y_3 = y_2 + h(23F_2 - 16F_1 + 5F_0)/12\,.$$

More generally, we have for n=2,3,...

$$y_{n+1} = y_n + h(23Fn - 16F_{n-1} + 5F_{n-2})/12\,. \tag{10.10}$$

The algorithm (10.10) is an example of *multi-step* method, so called because the expression for y_{n+1} makes use of more than just a single previous solution value. A single-step method uses only the prescribed initial condition to begin generating solution values, the computation of each new solution value making use of the solution value at the previous step only.

A multi-step method requires two or more "starting values" and usually makes use of a single-step method to generate them. When a sufficient number of solution values have been generated by the single-step method, the multi-step algorithm takes over. The algorithm in (10.10) is referred to as *three step* method because it requires solution values at three different levels, y_{n-2}, y_{n-1} and y_n. Several multi-step algorithms are listed here, together with the required starting values and the local truncation error for each algorithm:

Multi-Step Algorithms

A TWO-STEP METHOD: $E \sim Ch^3$

Starting Values: (x_0,y_0), (x_1,y_1)

$$y_{n+1} = y_{n-1} + 2hF_n \quad \text{for} \quad n=1,2,...$$

A THREE-STEP METHOD: $E \sim Ch^4$

Starting Values: (x_0,y_0), (x_1,y_1), (x_2,y_2)

$$y_{n+1} = y_n + h\,[23\,F_n - 16F_{n-1} + 5F_{n-2}] \quad \text{for} \quad n=2,\,3,\,...$$

A FOUR-STEP METHOD: $E \sim Ch^5$

Starting Values: (x_0,y_0), (x_1,y_1), (x_2,y_2), (x_3,y_3)

$$y_{n+1} = y_{n-3} + 4h\,[2\,F_n - F_{n-1} + 2\,F_{n-2}]/3 \quad \text{for} \quad n=3,4,\,...$$

A multi-step algorithm generally provides a higher order of accuracy than a comparable single step method. For example, the two-step method listed above is comparable to Euler's method in that each algorithm requires a single function evaluation per step. However, the local truncation error for the two-step method ($E \sim Ch^2$) is one order higher than for Euler's method ($E \sim Ch^3$). Generally speaking, this permits the two-step method to use a larger stepsize h than Euler's method in order to achieve equivalent accuracy.

In selecting a single-step method to generate the required starting values for a multi-step algorithm, care should be taken to use a method having the same order local truncation error as the multi-step method. Otherwise the accuracy advantage of the multi-step method may be lost.

SOLVED PROBLEMS

Euler's Method

PROBLEM 10.1

Prove the error estimates (10.4) from Theorem 10.1 .

SOLUTION 10.1

If y(x) solves (10.1) and F(x,y) has continuous derivatives of order one, then the differential equation implies that $y''(x)=\partial_x F(x,y)$ is continuous. We may then use Taylor's theorem to write

$$y(x_{n+1}) = y(x_n) + hy'(x_n) + y''(\xi)h^2/2 \quad \text{for} \quad x_n \le \xi < x_{n+1} \tag{1}$$

Then subtracting the equation

$$y_{n+1} = y_n + hF(x_n,y_n) \tag{2}$$

from (1), assuming that $y_n=y(x_n)$, leads to

$$E_{n+1} = |y''(\xi)h^2/2| \le Mh^2/2\,, \tag{3}$$

where M denotes the maximum value of $|\partial_x F(x,y)|$ on the closed set $a \le x \le b$, $A-Y \le y \le A+Y$, for sufficiently large constant Y.

If we subtract (2) from (1), assuming only that $y_0=y(x_0)$, then we obtain

$$e_{n+1} \le e_n + h\,|F(x_n,y(x_n)) - F(x_n,y_n)| + y''(\xi)h^2/2\,. \tag{4}$$

Now

$$|F(x_n,y(x_n)) - F(x_n,y_n)| = |\partial_y F(x_n,\upsilon)(y(x_n)-y_n)| \le Be_n, \tag{5}$$

where B denotes the maximum value of $|\partial_y F(x,y)|$ on the closed set $a \le x \le b$, $A-Y \le y \le A+Y$. Now, using (5) in (4), we find

$$e_{n+1} < e_n(1+Bh) + Mh^2/2 \quad \text{for} \quad n=0,1,... \tag{6}$$

If Δ_n, n=0,1,... denote a sequence of numbers satisfying

$$\Delta_n = 0 \quad \text{and} \quad \Delta_{n+1} = \Delta_n(1+Bh) + Mh^2/2\,,$$

then it is a simple exercise in mathematical induction to show that

$$e_n < \Delta_n \quad \text{and} \quad \Delta_n = \frac{Mh}{2B}((1+Bh)^n - 1) \quad \text{for} \quad n=0,1,... \tag{7}$$

Finally, since $1+Bh \le e^{Bh}$ and $Nh=L=b-a$, it follows from (7) that

$$e_n \le \frac{Mh}{2B}(e^{BL} - 1), \quad \text{for} \quad n=0,1,...$$

PROBLEM 10.2

Use Euler's method to compute y(1) if

$$y'(x) = 3x^2y^2 \qquad y(0)=1\,. \tag{1}$$

Carry out the computations for N=2,4,8,16, 32 and compare the results with the exact solution.

SOLUTION 10.2

The initial value problem (1) was solved in Problem 2.7, where the exact solution was found to be

$$y(x) = \frac{1}{1-x^3}\,.$$

For purposes of computing the Euler method approximation to the exact solution of this initial value problem, we have

$$h = (1 - 0)/N \quad \text{for the various choices of N.}$$

We shall show the steps in the computation in the case N=4 and display the results graphically for N=2,8,32.

Since $x_n=0+nh$, with $h=1/4$, we have $x_0=0$, $x_1=1/4$, $x_2=1/2$, $x_3=3/4$, and $x_4=1$. Then, according to (10.3),

$$y_0 = 1$$

$$y_1 = y_0 + 3x_0^2y_0^2h = 1 + 0 = 1$$

$$y_2 = y_1 + 3x_1^2y_1^2h = 1 + 3(.0625)1(.25) = 1.046875$$

$$y_3 = y_2 + 3x_2^2y_2^2h = 1.046875+3(.25)(1.0959472)(.25)=1.252365$$

$$y_4 = y_3 + 3x_3^2y_3^2h = 1.252365+3(.5625)(1.5684183)(.25)=1.914041$$

Note that although the exact solution becomes undefined at x=1, the approximate solution with N=4 assumes the value $y_4=1.914041$. As N increases and h=1/N decreases correspondingly, the approximate solution approaches the exact solution. The results with N=2,8, and 32 are graphed in Figure 10.2, together with the exact solution. As N increases, the graph of the approximate solution approaches the graph of the exact solution and the value of y_N grows. Of course, for every finite value of N, y_N will be finite.

The values of y_N for several values of N are as follows:

N	2	4	8	32
y_N	1.375	1.914	2.775	6.468

PROBLEM 10.3

Modify the algorithm for Euler's method so that it applies to the system of two equations,

$$\begin{aligned} dx/dt &= P(x,y,t) \qquad x(0)=A \\ dy/dt &= Q(x,y,t) \qquad y(0)=B. \end{aligned} \tag{1}$$

Write a computer program to carry out the computations for the algorithm.

SOLUTION 10.3

Using the notation $t_n=nh$ for n=0,1,... we can use Taylor's theorem to write

$$x(t_{n+1}) = x(t_n) + x'(t_n)h + \dots$$

$$y(t_{n+1}) = y(t_n) + y'(t_n)h + \dots$$

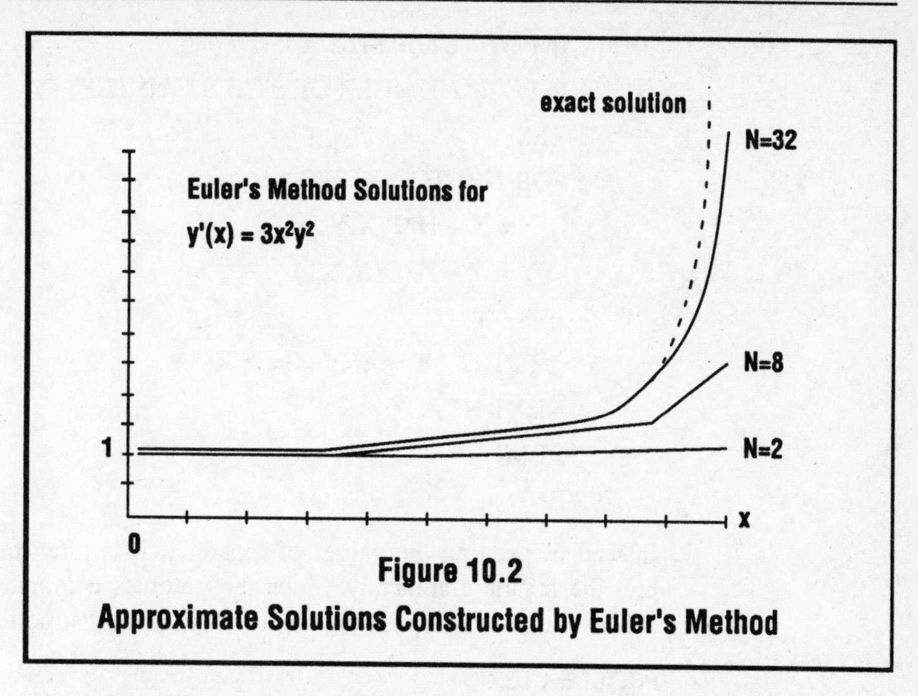

Figure 10.2
Approximate Solutions Constructed by Euler's Method

Now using (1) to replace the derivatives $x'(t_n)$ and $y'(t_n)$, we obtain

$$x(t_{n+1}) \sim x(t_n) + P(x(t_n),y(t_n), t_n)h$$

$$y(t_{n+1}) \sim y(t_n) + Q(x(t_n), y(t_n),t_n)h\,.$$

This leads to the following single step approximation scheme for the initial value problem (1)

$$x_0 = A \quad \text{and} \quad y_0 = B$$

$$x_{n+1} = x_n + P(x_n,y_n,t_n)h \qquad \text{for} \quad n=0,1, ...N-1 \tag{2}$$

$$y_{n+1} = y_n + Q(x_n,y_n,t_n)h\,, \quad \text{for} \quad n=0,1,... ,N-1\,.$$

The computations required by the algorithm (2) are best carried out on a computer. The following is a simple BASIC program for executing this algorithm in the case that $P(x,y,t)=y$ and $Q(x,y,t)=-2.5x$:

Two Dimensional Euler's Method Algorithm

```
REM 2-DIMENSIONAL EULER METHOD
REM
DEF P(X,Y,T) = Y
DEF Q(X,Y,T) = - 2.5 *X
REM
INPUT " INITIAL VALUES FOR T,X,Y "; T,X,Y
```

```
INPUT " STEP SIZE h "; H
INPUT " NUMBER OF STEPS "; NSTEPS
REM
FOR N=1 TO NSTEPS
  X = X + H*P(X,Y,T)
  Y = Y + H*Q(X,Y,T)
  T = T + H
  PRINT " T = "; T;" X = "; X;" Y = "; Y
NEXT N
REM
END
```

Instead of printing the values of t,x,and y, it is often more informative to plot the results graphically. Since the graphics commands even in BASIC are often machine dependent, we do not include that portion of the program.

PROBLEM 10.4

Use the algorithm from the previous problem to solve the initial value problem

$$\begin{aligned} x'(t) &= y(t) \qquad x(0) = 0 \\ y'(t) &= -b^2\, x(t) \qquad y(0) = A. \end{aligned} \tag{1}$$

Compute the solution and plot the corresponding orbit with the following values of step size: h=.02, .01, .005, .0025, .001.

SOLUTION 10.4

We have solved the initial value problem (1) in Problem 8.1, part (a), where we found that the orbit corresponding to this solution is the ellipse

$$b^2\, x^2 + y^2 = A\,. \tag{2}$$

Since $P(x,y,t)=y$ and $Q(x,y,t)=-b^2x$, the algorithm for this system reduces to

$$\begin{aligned} x_{n+1} &= x_n + y_n\, h \qquad x_0=0 \\ y_{n+1} &= y_n - b^2 x_n\, h \qquad y_0=A. \end{aligned} \tag{3}$$

Then

$$x_1 = 0 + Ah = Ah \qquad y_1 = A - 0 = A$$

$$x_2 = Ah + Ah = 2Ah \qquad y_2 = A - b^2Ah^2 = A(1-b^2h^2)$$

$$x_3 = 2Ah + Ah(1-b^2h^2) \qquad y_3 = A(1-b^2h^2) - b^2(2Ah^2)$$

etc.

Computing and plotting (x_n,y_n) for n=1,2,... should generate the elliptic orbit (2). Since the orbit is a closed curve, the point (x_n,y_n) begins to retrace the curve for n sufficiently large. However, the curve generated by the algorithm (3) is only an approximation to the true orbit.

For example, the curve generated by (3) with h=.02 is seen in Figure 10.3 to be a *spiral* rather than a closed curve. The orbits generated by (3) with successively smaller values for h are also shown in this figure. As the value for h is decreased, the value of the initial constant A was also decreased to allow the curves generated by different h values to be easily distinguished from one another. Note that as h is decreased from the value .02 to .01 and then to .005, the corresponding orbits can be seen to more nearly close upon themselves after completing one revolution. The orbit generated with h=.0025 is almost closed, only a thickening of the curve as the point (x_n,y_n) begins to retrace the ellipse indicates that the orbit is still slightly spiraling. Finally, with h=.001 the orbit is a closed curve with (x_n,y_n) retracing the same curve over and over. These results are consistent with Theorem 10.1, which predicts that for each n, the accumulated error, e_n, decreases to zero as h tends to zero.

Second Order Runge Kutta Method

PROBLEM 10.5

Show that the local truncation error for the Runge Kutta method of order two is proportional to h^3.

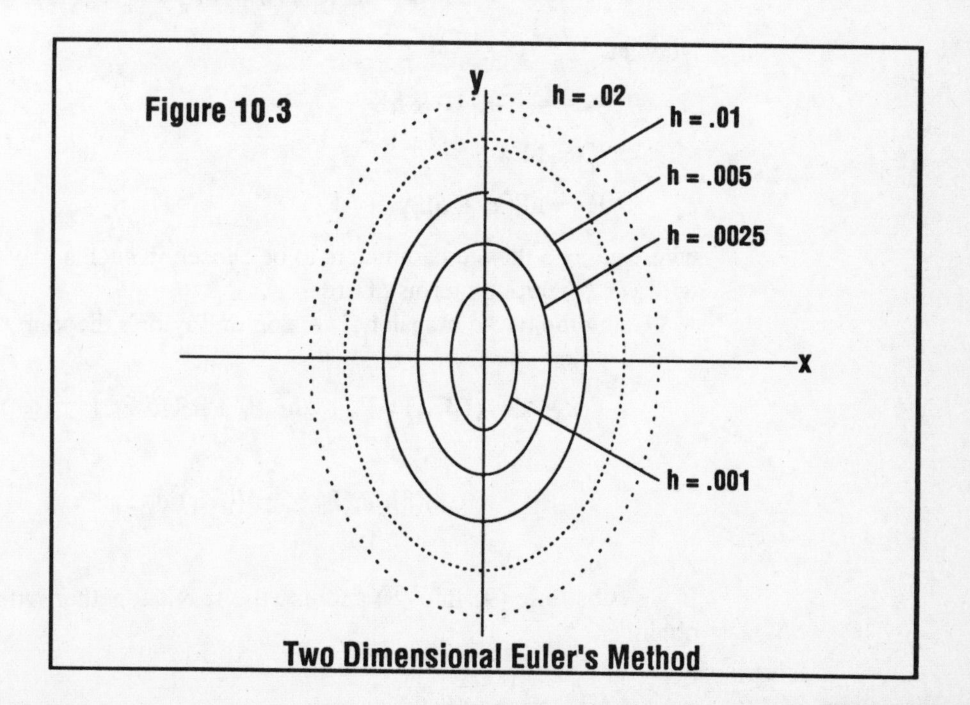

Figure 10.3

Two Dimensional Euler's Method

SOLUTION 10.5

Taylor's theorem implies that for y(x) sufficiently smooth,

$$y(x_{n+1}) = y(x_n) + hy'(x_n) + 1/2\, h^2\, y''(x_n) + 1/6\, h^3\, y^{(3)}(x_n) + \ldots \tag{1}$$

If y(x) satisfies (10.1) then

$$y'(x_n) = F(x_n, y(x_n)) \equiv F_n\,, \tag{2}$$

and if F(x,y) has continuous derivatives up to order two, then the chain rule implies

$$\begin{aligned} y''(x_n) &= \partial_x F(x_n, y(x_n)) + \partial_y F(x_n, y(x_n)) y'(x_n) \\ &= \partial_x F_n + \partial_y F_n \cdot F_n \end{aligned} \tag{3}$$

$$\begin{aligned} y^{(3)}(x_n) &= \partial_{xx} F_n + \partial_{xy} F_n \cdot F_n + \partial_x F_n \cdot \partial_y F_n \\ &\quad + F_n \cdot \partial_{yx} F_n + (\partial_y F_n)^2 F_n + F_n^2 \cdot \partial_{yy} F_n \end{aligned} \tag{4}$$

Substituting (2), (3), and (4) into (1) leads to

$$\begin{aligned} y(x_{n+1}) &= y(x_n) + hF_n + \frac{1}{2} h^2 [\partial_x F_n + \partial_y F_n \cdot F_n] \\ &\quad + \frac{1}{6} h^3 [\partial_{xx} F_n + \partial_{xy} F_n \cdot F_n + \partial_x F_n \cdot \partial_y F_n \\ &\quad + F_n \cdot \partial_{yx} F_n + (\partial_y F_n)^2 F_n + F_n^2 \cdot \partial_{yy} F_n] + \ldots \end{aligned} \tag{5}$$

Now let

$$y_{n+1} = y_n + aK_1 + bK_2 \tag{6}$$

$$K_1 = hF_n \tag{7}$$

$$K_2 = hF(x_n + \alpha h, y_n + \beta K_1) \tag{8}$$

where a, b, α, β denote parameters to be chosen in such a way that y_{n+1} agrees with $y(x_{n+1})$ through terms of order h^2.

Using the two-dimensional version of Taylor's theorem, we have

$$\begin{aligned} F(x_n+\alpha h, y_n+\beta K_1) &= F_n + \alpha h \partial_x F_n + \beta K_1 \partial_y F_n + \frac{1}{2}(\alpha h)^2\, \partial_{xx} F_n \\ &\quad + (\alpha h)(\beta h) \partial_{xy} F_n + \frac{1}{2}(\beta K_1)^2 \partial_{yy} F_n + \ldots \end{aligned} \tag{9}$$

If we substitute (9) into (8) and use the result together with (7) in (6), we obtain

$$y_{n+1} = y_n + (a+b)hF_n + (\alpha b\partial_x F_n + \beta bF_n\partial_x F_n)h^2 \qquad (10)$$
$$+ (\frac{1}{2} b\alpha^2\partial_{xx}F_n + \alpha\beta bF_n\partial_{xy}F_n + \frac{1}{2} b\beta^2F_n^2\partial_{yy}F_n)h^3 + \ldots$$

Then y_{n+1} given by (10) agrees with $y(x_{n+1})$ given by (5) through terms of order h if

$$a+b = 1, \quad \alpha b=1/2, \quad \beta b = 1/2 .$$

This set of three equations in four unknowns has infinitely many solutions. A convenient choice of solution is

$$\alpha = \beta = 1 \quad \text{and} \quad a = b = 1/2 .$$

Substituting these values into (6) through (8) produces the algorithm (10.5). Since y_{n+1} agrees with $y(x_{n+1})$ up to and including terms of order h^2, the local truncation error, $E_n = |y_{n+1} - y(x_{n+1})|$, is proportional to h^3.

PROBLEM 10.6

Use the Runge Kutta method of order two to compute y(1) if

$$y'(x) = 3x^2y^2 \qquad y(0)=1 . \qquad (1)$$

Carry out the computations with N=2,8, 32 and compare the results with the exact solution and with the approximate solutions generated by Euler's method.

SOLUTION 10.6

Recall that $x_n=nh$ for n=0,1,... N where h=1/N. Then $x_0=0$ and $x_N=1$. In addition, we are given $y_0=1$. Then in the case N=8, h=.125, the computation proceeds as follows,

$y_0 = 1$ $\qquad K_1 = 3x_0^2y_0^2h = 0$

$\qquad K_2 = 3x_1^2(y_0+K_1)^2 h = .0058$

$y_1 = y_0 + 1/2\,(K_1 + K_2) = 1.0176$ $\qquad K_1 = 3x_1^2y_1^2 h = .0058$

$\qquad K_2 = 3x_2^2(y_1+K_1)^2 h = .0236$

$y_2 = y_1 + 1/2\,(K_1 + K_2) = 1.057$ $\qquad K_1 = 3x_2^2y_2^2 h = .0242$

$\qquad K_2 = 3x_2^2(y_2+K_1)^2 h = .0549$

.

.

.

$y_7 = y_6 + 1/2\,(K_1 + K_2) = 4.180$ $\qquad K_1 = 3x_7^2y_7^2 h = 1.4988$

$\qquad K_2 = 3x_7^2(y_7+K_1)^2 h = 2.292$

These computations were carried out by computer and the results plotted in Figure 10.4 for N=2,8,32. Comparing this figure to Figure 10.2, it can be seen that for a given N, the Runge Kutta approximation to the solution of the initial value problem appears to be closer to the exact solution than the Euler approximation for the same N. This is a reflection of the fact that E_n, the local truncation error for this Runge Kutta method, is proportional to h^3, while the error for Euler's method is proportional to h^2.

PROBLEM 10.7

Modify the Runge Kutta algorithm of order 2 so that it applies to systems of the form

$$\begin{aligned} x'(t) &= P(x,y,t) \qquad x(0) = A \\ y'(t) &= Q(x,y,t) \qquad y(0) = B. \end{aligned} \tag{1}$$

Write a computer program to carry out the computations of the algorithm.

SOLUTION 10.7

The one-dimensional algorithm (10.5) approximates $y(x_{n+1})$ by the quantity

$$y_{n+1} = y_n + hF_n^{av}$$

where

$$F_n^{av} = \text{the average of } F_n \text{ and } F_{n+1}^{*} = F(x_{n+1}, y_n + hF_n)\,.$$

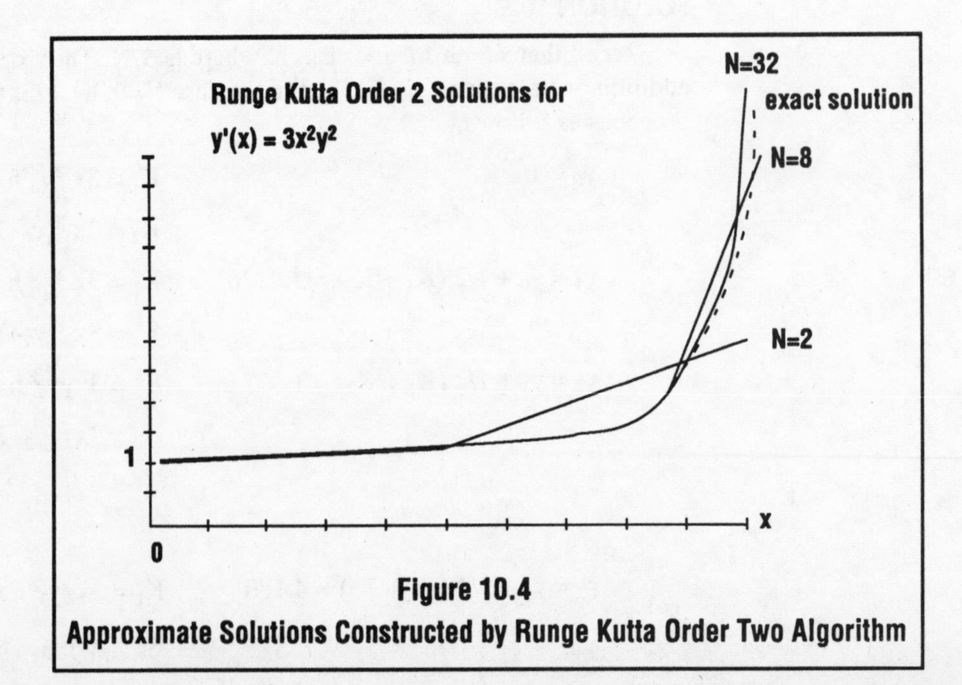

Figure 10.4
Approximate Solutions Constructed by Runge Kutta Order Two Algorithm

Note that F^*_{n+1} is the value that Euler's method would predict for F_{n+1}.

We can extend this algorithm to the two-dimensional system (1) by letting

$$XK1 = hP(x_n,y_n,t_n)$$

$$YK1 = hQ(x_n,y_n,t_n)$$

$$XK2 = hP(x_n+XK1,\ y_n+YK1,\ t_n+h)$$

$$YK2 = hQ(x_n+XK1,\ y_n+YK1,\ t_n+h)$$

Then

$$x_{n+1} = x_n + 1/2\ (XK1 + XK2)$$

$$y_{n+1} = y_n + 1/2\ (YK1 + YK2).$$

The following is a BASIC program to carry out the steps of this algorithm in the case that $P(x,y,t)=y$ and $Q(x,y,t)=-2.5x$:

Algorithm for Two Dimensional Runge Kutta (Second Order)

```
REM 2-DIMENSIONAL RK-2 METHOD
REM
DEF P(X,Y,T) = Y
DEF Q(X,Y,T) = - 2.5 *X
REM
INPUT " INITIAL VALUES FOR T,X,Y "; T,X,Y
INPUT " STEP SIZE h "; H
INPUT " NUMBER OF STEPS "; NSTEPS
REM
FOR N=1 TO NSTEPS
  XK1 = H*P(X,Y,T)
  YK1 = H*Q(X,Y,T)
  XK2 = H*P(X+XK1,Y +YK1, T+H)
  YK2 = H*Q(X+XK1,Y +YK1,T+H)
  X = X + (XK1 + XK2)/2
  Y = Y + (YK1 + YK2)/2
  T = T + H
  PRINT " T = "; T;" X = "; X;" Y = "; Y
NEXT N
REM
END
```

PROBLEM 10.8

Use the RK-2 algorithm from the previous problem to solve the initial value problem

$$\begin{aligned} x'(t) &= y(t) \qquad & x(0) &= 0, \\ y'(t) &= -b^2x(t) \qquad & y(0) &= A. \end{aligned} \tag{1}$$

Compute the solution and plot the corresponding orbit for h=.05, .01, .002.

SOLUTION 10.8

We have solved this problem using the two-dimensional Euler method in Problem 10.4. There we saw that for h=.02, .01, .005 the inaccuracy of Euler's method was sufficient to cause the orbit to form a spiral rather than the closed elliptical curve

$$b^2x^2 + y^2 = A^2,$$

which is the orbit for the exact solution. The step size h had to be decreased to .001 before the orbit generated by Euler's method appeared to be truly closed. In Figure 10.5 we have plotted the orbits generated by the Runge Kutta method of order two. There it can be seen that with step sizes roughly ten times as large as those used by Euler's method, the RK-2 algorithm achieves equivalent accuracy. In particular, the orbits generated by RK-2 with h=.2, .1 are spirals like the orbits generated by Euler's method for h=.02 and .01. For a step size of h =.05, the RK-2 algorithm generates a

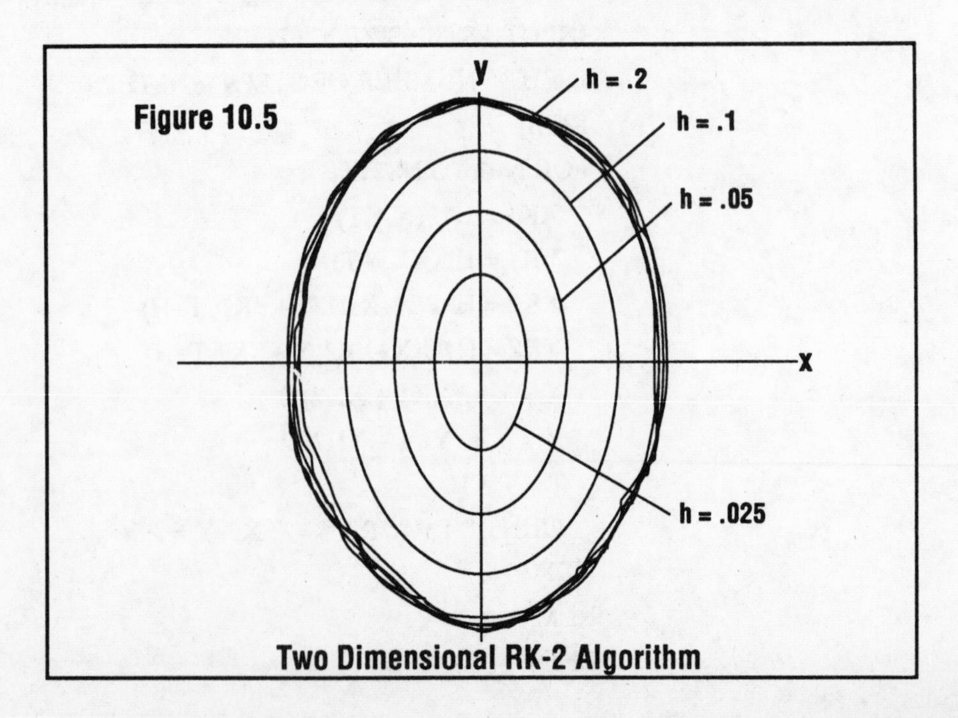

Figure 10.5

Two Dimensional RK-2 Algorithm

closed orbit while Euler's method requires a step size of h=.001 before the algorithm is sufficiently accurate for the orbit to be closed. Thus, while RK-2 requires that six lines of code be executed for each time step as opposed to two lines for the Euler's method, the RK-2 method allows step sizes at least ten times as large. Therefore, the RK-2 algorithm is more efficient.

Two Step Methods

PROBLEM 10.9

Show that the two-step algorithm

$$y_{n+1} = y_{n-1} + 2hF_n, \qquad n = 1, 2, 3, \dots, \tag{1}$$

approximates the solution of (10.1) and has local truncation error proportional to h^3.

SOLUTION 10.9

If y(x) solves (10.1), then y(0)=A, and

$$y(x_{n+1}) = y(x_{n-1}) + \int_{x_{n-1}}^{x_{n+1}} F(s,y(s))\,ds \tag{2}$$

Observe that the linear function

$$P(x) = [F_n(x-x_{n-1}) - F_{n-1}(x-x_n)]/h \tag{3}$$

passes through the points (x_{n-1},F_{n-1}) and (x_n,F_n), where for n=0,1,... N, x_n=nh and $F_n = F(x_n,y(x_n))$. In addition, it is easy to evaluate the integrals that show that P is such that

$$\int_{x_{n-1}}^{x_{n+1}} P(s)\,ds = F_n/h\int_0^{2h} z\,dz - F_{n-1}/h\int_{-h}^{h} z\,dz = 2hF_n\,.$$

Here we have split the integral into two parts and introduced the change of variable $z=x-x_{n-1}$ in the first integral and $z=x-x_n$ in the second. Then if we suppose (x_0,y_0), (x_1,y_1) are given, with y_0=A, and that $y_2,y_3,\dots$ are generated by

$$y_{n+1} = y_{n-1} + \int_{x_{n-1}}^{x_{n+1}} P(s)\,ds = y_{n-1} + 2hF_n \qquad n=1,2,\dots \tag{4}$$

it follows by subtracting (4) from (2) that

$$E_n = \left|\int_{x_{n-1}}^{x_{n+1}} [F(s,y(s)) - P(s)]\,ds\right| \tag{5}$$

It can be shown that

$$|F(s,y(s)) - P(s)| \le C|(x-x_{n-1})(x-x_n)| \quad \text{for} \quad x_{n-1} < s < x_{n+1}\,. \tag{6}$$

Then letting $z=x-x_n$ and using (6) in (5), we obtain

$$E_n \le \int_{-h}^{h} Cz(z+h)dz = 1/3\ Ch^3 .$$

The algorithm (1) requires starting values (x_0,y_0) and (x_1,y_1); these can be generated by the RK-2 algorithm. Once started, the algorithm (1) involves just one function evaluation per step as in Euler's method, but the truncation error is one order higher than for Euler's method.

PROBLEM 10.10

Use the two-point multi-step algorithm of the previous problem to approximate the solution of the initial value problem

$$y'(x) = 3x^2y^2, \qquad y(0) = 1, \tag{1}$$

on the interval (0,1). Carry out the computation for N=2,8,32 and compare with the exact solution and with the approximate solutions generated by Euler's method and the RK-2 algorithm.

SOLUTION 10.10

We require the starting values $(x_0,y_0)=(0,1)$ and $(x_1,y_1)=(h,1/(1-h^3))$. Then

$$y_2 = y_0 + 2hF_1 = y_0 + 2h(3x_1^2\, y_1^2)$$

$$y_3 = y_1 + 2hF_2 = y_1 + 2h(3x_2^2\, y_2^2)$$

etc.

The computations were carried out on the computer and the results plotted in Figure 10.6. The solutions produced by the two-step algorithm for N=2, 8, and 32 are roughly comparable to the solutions generated by the RK-2 algorithm for the same values of N. They are significantly better than the solutions generated by Euler's method for the same N. The two-step algorithm requires about the same effort per iteration as Euler's method (one function evaluation) and less effort than the RK-2 algorithm. However, the two-step algorithm requires one more starting value than either of the one-step algorithms.

PROBLEM 10.11

Show that the algorithm

$$y_{n+1} = y_{n-3} + 4h(2F_n - F_{n-1} + 2F_{n+1})/3, \qquad n=3,4,... \tag{1}$$

approximates the solution of (10.1) if the starting values (x_0,y_0), (x_1,y_1), (x_2,y_2) and (x_3,y_3) are given.

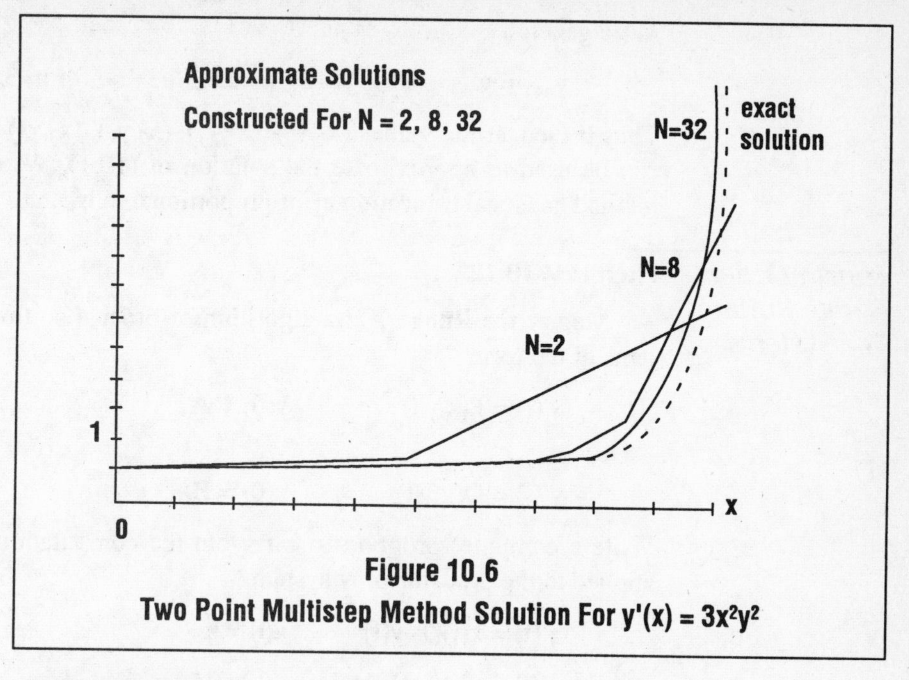

Figure 10.6
Two Point Multistep Method Solution For $y'(x) = 3x^2y^2$

SOLUTION 10.10

Note that the cubic function

$$P(x) = [F_3(x-x_0)(x-x_1)(x-x_2) - 3F_2(x-x_0)(x-x_1)(x-x_3) + 3F_1(x-x_0)(x-x_2)(x-x_3) - F_0(x-x_1)(x-x_2)(x-x_3)]/(6h^3) \tag{2}$$

passes through the points $(x_0,F_0),(x_1,F_1),(x_2,F_2),(x_3, F_3)$. In addition,

$$\int_{x_0}^{x_4} P(x)dx=[F_3\int_{-2h}^{2h} z(z+h)(z+2h)dz-3F_2\int_{-2h}^{2h}(z+h)(z-h)(z+2h)dz$$

$$+ 3F_1\int_{-2h}^{2h} z(z-h)(z+2h)dz - F_0\int_{-2h}^{2h} z(z-h)(z+h)dz]/(6h^3)$$

$$= h^4 (16F_3 - 8F_2 + 16F_1 - 0)/(6h^3)$$

$$= 4h(2F_3 - F_2 + 2F_1)/3 . \tag{3}$$

The solution of (10.1) satisfies

$$y(x_4) = y(x_0) + \int_{x_0}^{x_4} F(s,y(s))ds, \tag{4}$$

and since $P(x)$ approximates $F(x,y(x))$, we can approximate (4) by

$$y_4 = y_0 + \int_{x_0}^{x_4} P(s)\, ds = y_0 + 4h(2F_3 - F_2 + 2F_1)/3 .$$

More generally,

$$y_{n+1} = y_{n-3} + 4h(2F_n - F_{n-1} + 2F_{n-2})/3 \qquad \text{for } n=3, 4,...$$

Thus if the starting values (x_0,y_0), (x_1,y_1), (x_2,y_2), (x_3,y_3) are given, then (1) can be used to approximate the solution of (10.1). We can show that this method has local truncation error proportional to h^5.

Fourth Order Runge Kutta For Systems

PROBLEM 10.12

Modify the Runge Kutta algorithm of order 4 so that it applies to systems of the form

$$x'(t) = P(x,y,t) \qquad x(0) = A$$

(1)

$$y'(t) = Q(x,y,t) \qquad y(0) = B.$$

Write a computer program to carry out the computations of the algorithm applied to the predator-prey system

$$x'(t) = x(t)(1-y(t)) \qquad x(0)=A$$

$$y'(t) = y(t)(x(t)-1) \qquad y(0)=B. \tag{2}$$

SOLUTION 10.12

The one-dimensional RK-4 algorithm is given by (10.7). For systems of the form (1), it takes the form

$$x_0 = A \qquad y_0 = B,$$

$$XK_1 = hP(x_n,y_n,t_n)$$

$$YK_1 = hQ(x_n,y_n,t_n)$$

$$XK_2 = hP(x_n + XK_1/2, y_n + YK_1/2, t_n + h/2)$$

$$YK_2 = hQ(x_n + XK_1/2, y_n + YK_1/2, t_n + h/2)$$

$$XK_3 = hP(x_n + XK_2/2, y_n + YK_2/2, t_n + h/2)$$

$$YK_3 = hQ(x_n + XK_2/2, y_n + YK_2/2, t_n + h/2)$$

$$XK_4 = hP(x_n + XK_3, y_n + YK_3, t_n + h)$$

$$YK_4 = hQ(x_n + XK_3, y_n + YK_3, t_n + h)$$

$$x_{n+1} = x_n + (XK_1 + 2XK_2 + 2XK_3 + XK_4)/6$$

$$y_{n+1} = y_n + (YK_1 + 2YK_2 + 2YK_3 + YK_4)/6 \qquad \text{for } n=0,1, ...$$

The following is a BASIC computer program to carry out the steps of this algorithm applied to the predator-prey system (2) :

```
REM 2-DIMENSIONAL RK-4 METHOD
REM
DEF P(X,Y) = X(1–Y)
DEF Q(X,Y) = Y(X–1)
REM
INPUT " INITIAL VALUES FOR T,X,Y "; T,X,Y
INPUT " STEP SIZE h "; H
INPUT " NUMBER OF STEPS "; NSTEPS
REM
FOR N=1 TO NSTEPS
  XK1 = H*P(X,Y)
  YK1 = H*Q(X,Y)
  XK2 = H*P(X+XK1/2,Y +YK1/2 )
  YK2 = H*Q(X+XK1/2, Y +YK1/2 )
  XK3 = H*P(X+XK2/2,Y +YK2/2 )
  YK3 = H*Q(X+XK2/2, Y +YK2/2 )
  XK4 = H*P(X+XK3,Y +YK3 )
  YK4 = H*Q(X+XK3, Y +YK3 )
  X = X + (XK1 +2*XK2 + 2*XK3 + XK4)/6
  Y = Y + (YK1 + 2*YK2 + 2*YK3 + YK4)/6
  T = T + H
  PRINT " T = "; T;" X = "; X;" Y = "; Y
NEXT N
REM
END
```

This program was executed with h=0.1 for several choices of initial point (A,B). The pairs (A,B) were selected to produce distinct orbits, which have been plotted rather than printed and are displayed in Figure 10.7. These results are comparable to those shown in Figure 8.8. The orbits shown in Figure 8.8 were generated by a two-dimensional Euler's method with h= 0.001. Thus, the extra computational effort of the RK-4 algorithm allows the use of a step size that is 100 times as large as the one used for the simpler Euler's algorithm.

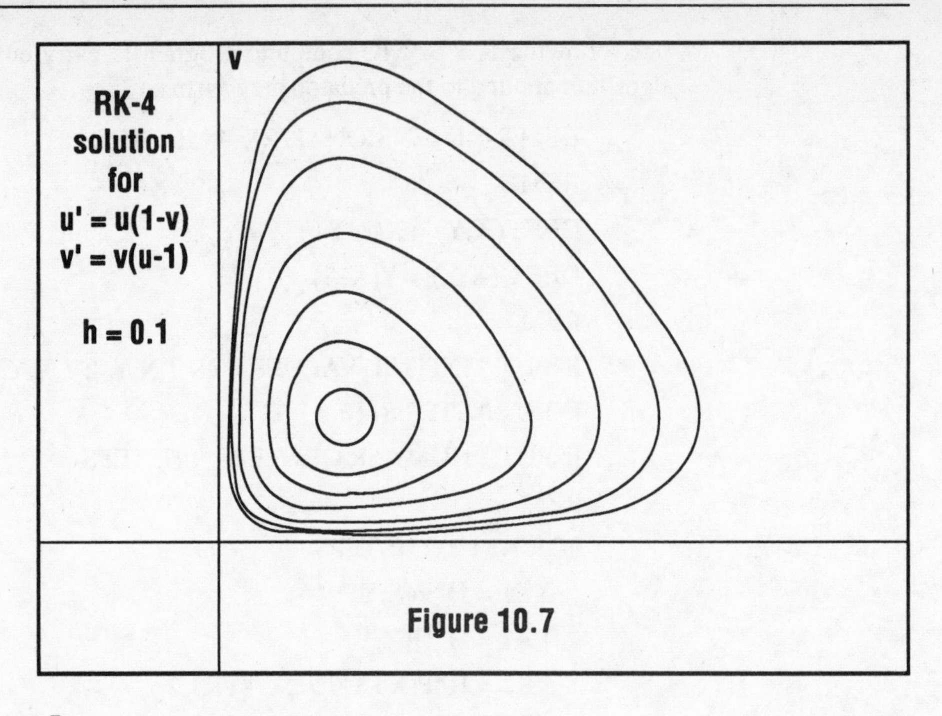

Figure 10.7

In this chapter we have briefly discussed means of constructing approximations to the solution of initial value problems of the form

$$y'(x) = F(x,y(x)) \qquad y(a)=A, \qquad \text{(one-dimensional)}$$

and

$$x'(t) = P(x(t),y(t),t) \qquad x(0)=A$$
$$y'(t) = Q(x(t),y(t),t) \qquad y(0)=B. \qquad \text{(two-dimensional)}$$

Single-step methods compute each new value of the approximate solution from a single previous value. Single-step methods discussed include:

Euler's method:	Algorithm (10.3)	one-dimensional	$E \sim h^2$
	Problem 10.3	two-dimensional	
Runge Kutta 2:	Algorithm (10.5)	one-dimensional	$E \sim h^3$
	Problem 10.7	two-dimensional	
Runge Kutta 4:	Algorithm (10.7)	one-dimensional	$E \sim h^5$
	Problem 10.12	two-dimensional	

For each method we have indicated that the local truncation error E is proportional to a power of the step size h. We refer to this power as the order of the truncation error. In general, the higher this order, the greater the accuracy of the method. In practical terms this means that in order to obtain a prescribed level of accuracy on a given problem, the RK-2 algo-

rithm could use a larger step size than Euler's method, and the RK-4 method could use an even larger step size. Large step size reduces computational effort.

Algorithms that require more than one previous solution value to compute the new value of the approximate solution are called multi-step methods. These algorithms generally make use of a single-step algorithm to generate the necessary starting values. Multi-step methods tend to be more accurate than single-step methods requiring comparable computational effort. However, this accuracy advantage can be lost if the order of the truncation error for the single-step starting method is not at least as high as for the multi-step method. Multi-step methods discussed in this chapter include:

(a) A Two-Step Method: $E \sim Ch^3$

Starting Values: (x_0,y_0), (x_1,y_1)

$$y_{n+1} = y_{n-1} + 2hF \quad \text{for} \quad n=1,2,...$$

(b) A Four-Step Method: $E \sim Ch$

Starting Values: (x_0,y_0), (x_1,y_1), (x_2,y_2), (x_3,y_3)

$$y_{n+1} = y_{n-3} + 4h\,(2\,F_n - F_{n-1} + 2F_{n-2})/3 \quad \text{for} \quad n=3,4, ...$$

Because a separate single-step method is required to generate starting values, the programming effort for computer implementation of multi-step methods is greater than for single-step methods.

Index

T

U

V

W

XYZ